Extended Consciousness and Predictive Processing

In this jointly authored book, Kirchhoff and Kiverstein defend the controversial thesis that phenomenal consciousness is realised by more than just the brain. They argue that the mechanisms and processes that realise phenomenal consciousness can at times extend across brain, body, and the social, material, and cultural world. Kirchhoff and Kiverstein offer a state-of-the-art tour of current arguments for and against extended consciousness. They aim to persuade you that it is possible to develop and defend the thesis of extended consciousness through the increasingly influential predictive processing theory developed in cognitive neuroscience. They show how predictive processing can be given a new reading as part of a third-wave account of the extended mind.

The third-wave claims that the boundaries of mind are not fixed and stable but fragile and hard-won, and always open to negotiation. It calls into question any separation of the biological from the social and cultural when thinking about the boundaries of the mind. Kirchhoff and Kiverstein show how this account of the mind finds support in predictive processing, leading them to a view of phenomenal consciousness as partially realised by patterns of cultural practice.

Michael D. Kirchhoff is senior lecturer in philosophy at the University of Wollongong, Australia.

Julian Kiverstein is senior researcher in philosophy at the University of Amsterdam, the Netherlands.

Routledge Focus on Philosophy

Routledge Focus on Philosophy is an exciting and innovative new series, capturing and disseminating some of the best and most exciting new research in philosophy in short book form. Peer reviewed and at a maximum of fifty thousand words shorter than the typical research monograph, *Routledge Focus on Philosophy* titles are available in both ebook and print on demand format. Tackling big topics in a digestible format the series opens up important philosophical research for a wider audience, and as such is invaluable reading for the scholar, researcher and student seeking to keep their finger on the pulse of the discipline. The series also reflects the growing interdisciplinarity within philosophy and will be of interest to those in related disciplines across the humanities and social sciences.

Political Theory and Global Climate Action
Recasting the Public Sphere
Idil Boran

Delusions and Beliefs
A Philosophical Inquiry
Kengo Miyazono

Émilie Du Châtelet and the Foundations of Physical Science
Katherine Brading

Extended Consciousness and Predictive Processing
A Third-Wave View
Michael D. Kirchhoff and Julian Kiverstein

What We Ought and What We Can
Alex King

For more information about this series, please visit: www.routledge.com/Routledge-Focus-on-Philosophy/book-series/RFP

Extended Consciousness and Predictive Processing

A Third-Wave View

Michael D. Kirchhoff and Julian Kiverstein

Routledge
Taylor & Francis Group
LONDON AND NEW YORK

First published 2019
by Routledge
2 Park Square, Milton Park, Abingdon, Oxon OX14 4RN

and by Routledge
52 Vanderbilt Avenue, New York, NY 10017

Routledge is an imprint of the Taylor & Francis Group, an informa business

British Library Cataloguing-in-Publication Data
A catalogue record for this book is available from the British Library

Library of Congress Cataloging-in-Publication Data
Names: Kirchhoff, Michael D., author.
Title: Extended consciousness and predictive processing : a third wave view / Michael D. Kirchhoff and Julian Kiverstein.
Description: 1 [edition]. | New York : Routledge, 2018. | Series: Routledge focus on philosophy | Includes bibliographical references and index.
Identifiers: LCCN 2018045588 | ISBN 9781138556812 (hardback : alk. paper) | ISBN 9781315150420 (e-book)
Subjects: LCSH: Philosophy of mind. | Cognitive science. | Consciousness.
Classification: LCC BD418.3 .K57 2018 | DDC 128/.2—dc23
LC record available at https://lccn.loc.gov/2018045588

ISBN: 978-1-138-55681-2 (hbk)
ISBN: 978-1-315-15042-0 (ebk)

Typeset in Times New Roman
by Apex CoVantage, LLC

Contents

Boxes, figures, and tables

Boxes

Figures

Tables

Acknowledgements

Many people have been involved in cultivating the ideas in this book. We would especially like to thank John Sutton, Richard Menary, Andy Clark, Mike Wheeler, Karl Friston, Jakob Hohwy, Evan Thompson, Alva Noë, Dan Zahavi, Shaun Gallagher, Andreas Roepstorff, Daniel. D. Hutto, Erik Myin, Regina Fabry, Erik Rietveld, Ludger Van Dijk, Jelle Bruineberg, Thomas Parr, Maxwell Ramstead, Ensor Palacios, Micah Allen, Somogy Varga, Krzysztof Dolega, Ian Robertson, Anco Peeters, Tom Froese, Valtteri Arstila, Joseph Almog, Dave Ward, Giovanna Colombetti, and finally Susan Hurley, whose work continues to prove a rich source of inspiration. Closer to home, M.K. would like to thank Casey Gibson, Billy, and Evig Glad (aka Forever Happy). J.K. would like to thank Catriona Black and his children Calum and Anna Kiverstein. We would also like to extend our gratitude for funding support for this project to the Australian Research Council (M.K. is funded by an Australian Research Council Discovery Project (DP170102987)), the John Templeton Foundation (M.K. is funded by a John Templeton Foundation Academic Cross-Training Fellowship (ID no. 60708)), and the European Research Council (J.K. is funded by the European Research Council in the form of a starting grant (679190) awarded to Prof. Erik Rietveld). Finally, we would like to thank the team at Routledge and all the anonymous reviewers involved in improving the quality of this book.

Some of the material in this book is motivated by previously published or forthcoming work. Chapter 1 expands on material from the following articles: Kirchhoff, M.D. (2014). Extended Cognition & Constitution: Re-evaluating the Constitutive Claim of Extended Cognition. *Philosophical Psychology*, 27(2), 258–283 and Kiverstein, J. (2017). Extended Cognition. In Gallagher, S., Newen, A., and De Bruin, L. (eds), *Oxford Handbook of Embodied, Enactive and Extended Cognition*. Oxford University Press. Chapter 3 builds on material from Kirchhoff, M.D. (2017). Predictive Processing, Perceiving, and Imagining: Is to Perceive to Imagine, or Something

Close to It? *Philosophical Studies*, 1–17. DOI: 10.1007/s11098-017-0891-8, and Bruineberg, J, Kiverstein, J., and Rietveld, E. (2016). The Anticipating Brain is Not a Scientist (Synthese SI: Predictive Brains). Chapter 4 comprises entirely new material. However, the simulations and figures are based on Kirchhoff, M.D., Parr, T., Palacios, E., Friston, K.J., and Kiverstein, J. (2018). The Markov Blankets of Life: Autonomy, Active Inference and the Free Energy Principle. *The Journal of the Royal Society Interface*, 15, 20170792, and Palacios, E., Razi, A., Parr, T., Kirchhoff, M.D., and Friston, K. (2017). Biological Self-Organisation and Markov Blankets. *BioRxiv*, DOI: http://dx.doi.org/10.1101/227181. Chapter 6 builds on material from Kirchhoff, M.D. (2015c). Species of Realisation and the Free Energy Principle. *Australasian Journal of Philosophy*, 93:4, 706–723 and Kirchhoff, M.D. (2015a). Extended Cognition & the Causal-Constitutive Fallacy: In Search for a Diachronic and Dynamical Conception of Constitution. *Philosophy and Phenomenological Research*, 90(2), 320–360.

Introduction

In this jointly authored book, we will argue that perceptual consciousness is sometimes realised by cycles of embodied and world-involving engagement. Thus, the realisers of perceptual experience can extend beyond the brain to include bodily and worldly elements. We call this the *thesis of extended consciousness*. We do not claim this is true of all forms of conscious experience. The kind of experiences we have while imagining something, or during dreaming, are forms of experience that are less directly tied to bodily activity than the experiences of the world we have during online perception. The position we defend here is that the mind has no fixed boundary. The locus of conscious experience can smoothly shift from on occasions being inside of the head of the individual to on other occasions forming out of a nexus of interactions between brain, body, and environment.

We will argue for the thesis of extended consciousness through the increasingly powerful predictive processing theory developed in cognitive neuroscience (Clark 2013; Friston 2010). The predictive processing theory claims that minds in general, starting from those of the simplest of living organisms and scaling up to the human mind, work according to a single imperative to minimise prediction error. Living beings develop prior expectations about the sensory observations they harvest from the world. The organism then puts its prior expectations to work to proactively predict the flow of sensory information that arises from its engagement with the world. The imperative to minimise prediction error is nature's way of optimising the evidence for the prior expectations that inform its predictions (Clark 2016; Friston 2011; Hohwy 2013; Kirchhoff 2017; Seth 2015a).

Predictive processing is standardly taken to imply a view of perceptual consciousness as a process of probabilistic inference. In their orthodox formulations, predictive processing theories take conscious experience to be the result of neural activity, over different temporal and spatial scales, that best minimises prediction error (Clark 2016; Friston et al. 2018; Hohwy 2013; Seth 2015a). As Hohwy (2012) observes, "what gets selected for conscious perception is the hypothesis or model that, given the widest context, is currently most closely guided by the current (precise) prediction errors" (2012, p. 5). This is a somewhat deflationary account of consciousness in which conscious experience just is the hypothesis that best accounts for current sensory evidence. We add two important twists to this story.

The first twist is that we take predictive processing to be in the business of regulating how we actively sample our environment and therefore proactively orchestrate our own sensations (Friston et al. 2018). This implies that predictive processing is not a theory of unconscious *perceptual* inference (c.f. Gregory 1980; Helmholtz 1860). It is a theory of how organisms engage in embodied activity with the aim of maintaining themselves within certain expected states (Bruineberg et al. 2016; Kirchhoff et al. 2018). Our account of predictive processing starts from what is called "active inference." Organisms act with the aim of keeping to a minimum time spent in unexpected or surprising sensory states (Friston and Stephan 2007). Under active inference, conscious experience has its roots in embodied activity, coupling the agent – that is, the generative model – to the world. Online perceptual experience – but not necessarily the kind of experience we have when dreaming or imagining – is constituted in what we, after Susan Hurley (1998), call an extended dynamic singularity – a continuous network with internal and external feedback loops.

The second twist is that we frame this prediction-driven view of conscious experience through the prism of a third-wave account of the extended mind (Sutton 2010; see also Kirchhoff 2012).[1] The central tenets of third-wave extended mind detailed in Table 0.1 inform our interpretation of the philosophical implications of predictive processing.

In the remainder of our introduction, we provide a brief overview of each of the chapters to follow. Chapter 1 introduces the third wave in the extended mind debate by comparing it to its first-wave and second-wave cousins.

Table 0.1 Key tenets of third-wave extended mind

1. *Extended Dynamic Singularities*: some cognitive processes are constituted by causal networks with internal and external orbits comprising a singular cognitive system.
2. *Flexible and Open-Ended Boundaries*: the boundaries of mind are not fixed and stable but fragile and hard-won, and always open to negotiation.
3. *Distributed Cognitive Assembly*: the task and context-sensitive assembly of cognitive systems is driven not by the individual agent but by a nexus of constraints, some neural, some bodily, and some environmental (cultural, social, material).
4. *Diachronic Constitution*: Cognition is intrinsically temporal and dynamical, unfolding over different but interacting temporal scales of behaviour.[2]

Chapter 2 makes an argument for extended consciousness based on ideas drawn from sensorimotor enactivism (c.f. Hurley 1998). We argue that perceptual consciousness is realised in interactions among neural, bodily, and environmental dynamics and that these interactions have a particular temporal structure. This argument is referred to as the DEUTS (dynamic entanglement and unique temporal signature) argument for extended consciousness.

Clark has attempted to block this route to an argument for extended consciousness, in part based on considerations drawn from his work on predictive processing (Clark 2009, 2012). He has suggested that predictive processing leads to a view of perceptual experience as a form of "controlled hallucination." The brain uses the sensory signal to test its internally generated best guess about what might be going on in the world. The same processes of probabilistic guessing that take place in waking experience can also be used in imagining and dreaming. The only difference is that imagining and dreaming are modes of predictive processing free from the guidance of the incoming sensory signal. Phenomenologically, however, perceiving, dreaming, and imagining share much in common, because all three types of experience are generated by the brain with interaction with the world, serving only to constrain its probabilistic inferences. Chapter 3 develops a response to Clark (and others that have made similar arguments). We show how perceptual experience of the world is constitutively dependent on an extended dynamic singularity. Offline conscious experience in dreaming or in imagining is possible given that the dynamics of agent-environment relations are recycled during offline activity.

Chapter 4 takes as its starting point a recent argument against extended consciousness made by Chalmers. Chalmers proposes to use perception and action as a criterion for demarcating the boundaries of the mind. The intuitive idea is that the mind begins where the senses meet the world and ends when motor action brings us into contact with things. Perception and action can thus be thought of as interfaces where mind meets world. Hohwy (2016, 2017a) has argued along similar lines drawing on predictive processing to mount a more general attack on the extended mind. Chapter 4 develops a response to Chalmers and Hohwy by showing that there is no single, unique, fixed boundary that serves to demarcate the mind from the rest of the world.

Chapter 5 is concerned with the role that prior expectations play in shaping conscious experience. Predictive processing claims that conscious experience is the outcome of settling on an interpretation of how things are in the world that does the best job of reducing prediction error across the sensory hierarchy as a whole. This requires balancing the influence of top-down expectations with incoming sensory information. In the predictive processing theory, the work of modulating the influence of prediction error on downstream processing is done by the precision mechanism. We show how the patterned regularities of cultural practices can make a functional contribution to weighing the precision of prediction error signals. Many of the expectations that come to guide perception and action are weighted as highly reliable because they align with the expectations of other people that take part in a given practice. The result of allowing precision to be set in part by cultural practices is that people come to share culturally specific ways of perceiving the world. Their experiences come to be shaped by the cultural niches they construct. Through the argument from distributed assembly, we thereby show how cultural practice can play a part in the constitution of conscious experience.

Finally, Chapter 6 argues for a diachronic perspective on the constitution of consciousness. The received view in metaphysics is that constitution is a synchronic relation that holds at a durationless instant *t* – a point that serves to distinguish constitution, say, from causation (Bennett 2011). We argue that conscious experience is a process and that the constitutive basis of consciousness is diachronic. We show how such a diachronic view of constitution is mandated by taking into account a number of key properties of the predictive processing framework when addressing consciousness. Our final chapter thus serves two purposes: to speak to wider issues in the metaphysics of constitution and dependence and to further develop our account of phenomenal consciousness as constituted by embodied engagement with the world.

Before we bring our introduction to a close, we provide some brief remarks concerning what we do not attempt in the book. In what follows, we do not attempt to explain how it is that predictive processing metaphysically necessitates phenomenal consciousness. We do not discuss the hard problem of consciousness (c.f. Chalmers 1995). Nor do we address the arguments that have been made for an explanatory gap (c.f. Levine 1983). Instead, we follow Hurley's (2010) useful distinction between her two terms "how-explanations" and "what-explanations." How-explanations aim to identify the processes and mechanisms that explain some phenomenon of interest, such as phenomenal consciousness. What-explanations aim to account for what a subject experiences. We will use predictive processing to develop a how-explanation of phenomenal consciousness. We show how the resulting explanation is one that allows the realisers of phenomenal consciousness to sometimes extend into the environment. We look to sensorimotor enactivism for what-explanations. It is sensorimotor enactivism which properly understood can be shown to entail an extended view of phenomenal consciousness.

The how- and what-explanations we develop do not speak directly to the questions people raise when they worry about the hard problem. Our book will therefore not satisfy readers who worry about why it should be that creatures who are functionally and physically like us should enjoy any conscious mental life whatsoever. We simply take it for granted that the world is populated with conscious creatures, and we do not attempt to resolve whatever people have found to be mysterious about this fact. Our question is *where* consciousness happens when it happens, not the question of why it happens at all. We argue that the answer to this question can vary over time. The boundaries of the conscious mind are fluid and negotiable.

Notes

1 John Sutton (2010) initiated the call for a third wave to the extended mind.
2 We note here that the first and final theme of the third wave are not explicitly part of Sutton's (2010) original formulation of the tenets of the third-wave view of the extended mind. We think, however, that they follow from taking seriously the implications of thinking about mentality from the perspective of the third-wave view of the extended mind.

1 The extended mind

Three waves

1.1 Introduction

The topic of this chapter is the debate surrounding the boundaries of the mind. We will be concerned with the thesis discussed in the literature on the extended mind that the mechanisms that produce cognitive processes are not always located within the individual organism. We use the term "cognitive processes" to refer to the types of processes that are investigated in the cognitive sciences, broadly construed. Examples include "higher cognitive processes" such as cognitive control, memory, decision-making, social cognition, mathematical reasoning, and language understanding. We will take the term "cognition" to also encompass examples of what some would call "lower" or "basic" cognitive processes, such as sensorimotor coordination.

Defenders of the extended mind oppose a position we will label "internalism," a term that covers a variety of different views. Metaphysical internalism is the view that the mechanisms that realise a given psychological capacity are located inside of individual agents. In the case of biological agents, the changes in patterns of neural activation found within the brain of an individual are metaphysically sufficient for some cognitive process of interest (Adams and Aizawa 2008; Aizawa and Gillett 2009a, 2009b; Bickle 2003; Churchland 1992; Fodor 1975; Gillett 2002, 2003, 2007; Hohwy 2013, 2016; Kim 1998).

The other form of internalism that will interest us is methodological. This is the view that the unit of psychological analysis should be the individual (Rupert 2009, 2004; and all references above). Methodological internalists are thus especially interested in the internal operations of certain specific mechanisms underlying cognitive behaviour – for example, the connection between Wernicke and Broca regions in the brain and how they are involved in speech production and its various pathologies. Or, they might ask if there a particular network in the brain – for example, the default network that underlies our ability to conceive of ourselves and others as bearers of mental states. In philosophy of mind, methodological internalism is sometimes referred to as individualism (Bechtel 2009; Burge 1979). Individualism is the thesis that any kind of behaviour is to be explained in terms of psychological processes conceived of as properties of individuals taken in isolation from their (social) environments.

Internalism can be questioned. Methodological internalism has been questioned since the 1970s, which saw the rise of content externalism. Content externalism is the position that meaning or the content of mental states depends constitutively on the kind of environment in which individuals are situated (Burge 1986; Putnam 1975). Content externalists therefore question the assumption that the content of our thoughts is determined solely by internal matters of fact. Appearances to the contrary, content externalism is in fact consistent with metaphysical and methodological internalism. It is silent about the location of the mechanisms and processes that realise an individual's psychological capacities.

Active externalism rejects both metaphysical and methodological internalism. This form of externalism saw its first articulation in 1998 in Andy Clark and David Chalmers's "The Extended Mind" – though it should be noted that there are important precursors in, for instance, the work of Edwin Hutchins (1995) and Robert Wilson (1994). Active externalism is "active" in the sense of allowing elements of the environment to play "an active role in driving cognitive processes" (Clark and Chalmers 1998, p. 7). The "driving influence" they had in mind was not the uncontroversial claim that elements of the external environment causally influence cognitive processes or mental states. As already noted in our introduction, even the most hardened internalist will allow this much of a role for the environment. Clark and Chalmers were out to defend the altogether more daring and controversial thesis that elements located in the environment are sometimes literally parts of our mind – proper parts of the conditions enabling us to perform certain cognitive processes or tasks (Clark 2008; Menary 2007; Rowlands 1999; Sutton 2010; Wheeler 2010).

A familiar example is the relationship people have with their mobile phones. Barely a moment passes when we find ourselves alone before we reach into our pockets for our phones. Chalmers (2008) tells us that his mobile phone has replaced part of his memory. He uses it to store information about his favourite restaurants and the dishes they serve. He uses it to do different calculations that would otherwise tax his onboard, brain-based memory capabilities. Part of the work of remembering that was previously done by Chalmers's brain is now done through the combined effort of his brain and his phone. This is an example of an extended cognitive process – memory is partially realised in a technological device located outside of the head of the individual.

In the remainder of this chapter, we set aside questions about consciousness, for the extended mind (EM) was originally developed as limited in its scope to unconscious cognitive processes and dispositional mental states. Its generalisation to consciousness is resisted by its principal architects (see, e.g., Clark 2009; Wheeler 2015a, b).

EM claims that under certain conditions cognitive processes and cognitive systems may be *constituted* or *realised* by resources distributed across the brain, the body, and the environment (Clark and Chalmers 1998; Clark 2008; Heersmink 2014, 2016; Kirchhoff 2012; Menary 2007, 2010; Rowlands 1999; Sutton 2010; Wheeler 2010, 2015b). It is thus a thesis that concerns the nature and breadth of the constituting base of cognition. We use the term "constitution" to refer to the metaphysical relation of dependence that holds between causal mechanisms and processes (typically biological) and the cognitive process whose nature we are interested in explaining. In sharp contrast to internalism of any stripe, EM is the view that in addition to neural activity within an individual, extra-neural bodily and environmental elements may be parts of the metaphysically sufficient conditions for some forms of cognition.[1]

Internalist theories share in common the claim that while elements of the environment may exert a causal influence on cognition, such elements cannot contribute to the constitution or realisation of a cognitive process. Internalists maintain that cognitive processes are physically constituted by processes housed within the skull of the individual. Opponents of EM often accuse its advocates of failing to observe the metaphysical difference between a relation of mere causal dependence and a relation of metaphysical constitution or realisation (c.f. Adams and Aizawa 2008). We will have ample opportunity to repeatedly return to this vexed issue. We set this issue aside for now in favour of further developing the details of EM. While we intend to defend a EM position in what follows, our aim in this chapter is to get clear on the commitments of EM.

Within EM, three different waves or ways of articulating EM have emerged, each of which endorse different perspectives on the nature of these conditions. In the remainder of this chapter, we survey these different waves in the extended cognition debate. Our aim is to showcase the theoretical virtues of a third-wave approach to the extended mind (henceforth "third-wave extended mind").

1.2 First-wave extended mind

First-wave EM is good old-fashioned role functionalism, associated either with common-sense functionalism (Clark and Chalmers 1998; Clark 2008) or with psychofunctionalism (Wheeler 2010). Functionalists claim that what different token mental states of the same type have in common is the causal role they occupy. Thus, a belief that it is raining is a state that is typically caused by perceptual experiences of rain and when taken together with a desire not to get wet causes the action of leaving the house with an umbrella. Functionalists typically allow that mental states of the same type can be physically realised in different substrates – this is known as multiple realisability. EM is argued to follow from a commitment to the multiple realisability of the mental (Wheeler 2010). Action on external information-bearing structures is argued to be one of the many ways in which cognitive processes are realised in the world we live in (Wilson 2004).

The causal role that defines a state as one of remembering (to take one much-discussed example from the EM literature) can be realised neurally but can also be realised in processes that combine neural, bodily, and environmental structures. Simple examples include a shopping list and doing calculations with pen and paper (Figure 1.1).

When going shopping, say, one could remember all the items on the list internally relying on something like a short-term memory buffer. Alternatively, one could write down the list of things one needs to buy and consult the list as one shops for groceries. According to EM, we should count these two possibilities as different physical implementations of the same process of remembering. In one case, information is temporarily stored internally; in the other case, it is temporarily stored in the world. We should

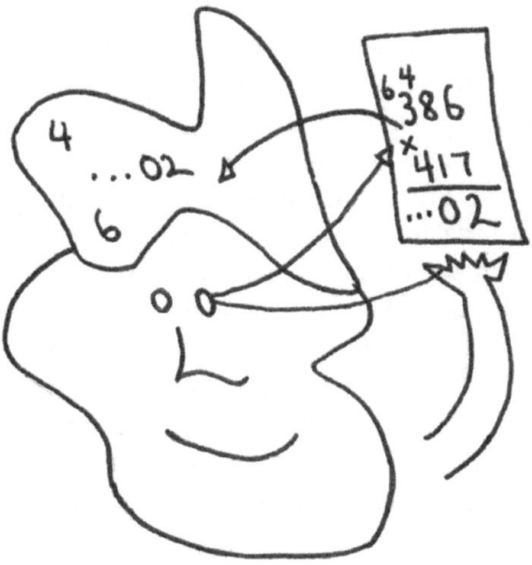

Figure 1.1 Incorporation of external elements into a process of calculation

Source: adapted from Wilson and Clark 2009, p. 10; used with permission

not treat the two cases differently simply on the basis of the location of the information-bearing structures. A process ought to be accorded the status of memory on the basis of the causal role that it plays in the guidance and control of behaviour, not based on the location of the process inside or outside of the head. The line of argument we have just sketched rests on what Clark and Chalmers called the parity principle. This principle states:

> If, as we confront some task, a part of the world functions as a process which, were it to go on in the head, we would have no hesitation in accepting as part of the cognitive process, then that part of the world is (for that time) part of the cognitive process.
>
> (1998, p. 7)

The parity principle can be thought of as a heuristic for establishing whether a putative world-involving process counts as a case of extended cognition. First, we identify an "external" element taken to perform a functional role in intelligent behaviour. Second, we imagine a scenario in which the same functional role is realised by an "internal" mechanism. Finally, we ask the question, should we count the "internal" mechanism as a part of the cognitive system? If the answer is "yes," then by parity of reasoning, we should say the same of the externally located element (e.g., the shopping list). We should not treat it differently simply on the basis of its location outside of the head.[2]

We use the term "parity-based arguments" for arguments for EM based on the parity principle. The soundness of parity-based arguments depends on judgements of functional similarity of internally housed and environmentally located elements. They require us to evaluate whether an element located in the external environment could be said to be performing the same functional or causal role as an internal mechanism located inside of an individual's head. Proponents of EM have typically favoured relatively coarse-grained criteria for individuating causal roles (Clark 2008). Thus, they have suggested that elements count as contributing to the same causal role if they play the same role in the guidance of action. Opponents of EM have tended to favour more fine-grained criteria (Rupert 2009; Adams and Aizawa 2010). They have pointed to significant functional differences in remembering items using a list and biological memory. They argue that such a functional difference militates against treating extended and internal cognitive processes as different realisations or occupants of the same functional or causal role.

A sticking point in the debate has thus been settling on the right criteria for specifying a causal role (Rowlands 2009). This is of course the long-standing problem of the inputs and outputs, as identified by Block (1978) in his early critique of functionalism. If we stick too closely to the fine-grained details of the functional profile of humans, we end up reaching chauvinist verdicts when it comes to the mentality of creatures that do not share our functional profile. If, on the other hand, we allow for a more abstract characterisation of functional similarity, we end up being too liberal. We have to allow systems like the economy of Bolivia to count as functionally similar to us. The point of controversy has been to find the right balance, and this continues to be a problem that plagues the debate surrounding EM (see, e.g., Sprevak 2009 for detailed discussion).

1.3 Second-wave extended mind

The second-wave formulation of EM stresses functional dissimilarities, inter alia, between internal and external elements. While the first-wave debate centred on parity-based arguments for EM, the second-wave debate aims to downplay and is often critical of parity considerations (Menary 2010; Sutton 2010). It aims to provide a different set of considerations in support of EM that do not turn on the commitments of one's preferred version of functionalism.

Second-wave theories argue that external structures like shopping lists complement the capacities of the biological brain precisely because of their different functional properties. First-wave EM requires a highly abstract, coarse-grained functional equivalence of internal and external elements as tested for by parity considerations. In doing so, it encourages us to focus on the similarity of internal and external components and therefore to downplay any putative differences between extra-neural and neural operations. It does so of course to call into question entrenched beliefs about the boundaries of the mind. The parity principle invites us to set aside prejudices that would lead us to rule out in advance items located outside of the boundaries of the biological individual from occupying a place in the realm of the cognitive.

In contrast to first-wave EM, second-wave EM "both predicts and requires" (Rowlands 2010, p. 89) a variety of ways in which "non-biological scaffolding" can augment and complement the brain's biological modes of processing (Sutton 2010, p. 189). It stresses different but complementary functional properties, whereas the first wave insists on similarity in functional properties. Props and artefacts have different formats: dynamical properties and functions as compared to internal, biological states and processes. Yet they can be brought together through agent-environment couplings to make complementary but heterogeneous contributions in the performance of a cognitive task. The resulting hybrid cognitive systems are part biological and part technological, as well as cultural (Malafouris

2010). Moreover, they are genuinely transformatory of our cognitive capacities, yielding new cognitive powers that we would otherwise lack precisely because of the different functional properties of the external, non-biological scaffolding (Menary 2007). The latter delivers something genuinely new that could not be accomplished by relying on only internal biological formats of processing. It is these different functional properties that the second wave claims a science of extended cognition ought to be interested in (Sutton 2010). This means studying the particular fine details of the affordances of the artefacts that are employed in cognition and their often particular and idiosyncratic histories of use (Malafouris 2004).

In second-wave EM, two different but overlapping formulations can be distinguished. The first turns on a complementarity principle (Sutton 2010, p. 205), while the second is more "enactive" in its themes, stressing integration and manipulation (Rowlands 1999; Menary 2007, 2010). Sutton's complementarity approach stresses the "wildly heterogeneous" and the various dimensions of difference in the nature of the components that make up an extended cognitive system (see also Heersmink 2016). The brain interfaces with external resources, such as media, institutions, and other people, rather than replicating internally the capacities that such resources make available. External resources complement the "brain's style of storage and computation" (Clark 1997, p. 220) but differ from brain processes in terms of their format and dynamics. Thus, in Sutton and colleagues' work on memory, for instance, the functional differences between engrams (intracranial, biological forms of information preservation) and exograms (extracranial, symbolic forms of information preservation) are stressed. Engrams rely on a distributed, superposed, and context-dependent reconstruction of information over time. Exograms, such as written notes, by contrast, exploit a discrete and symbolic medium that will generally allow for a stable, context-independent, and relatively enduring forms of information storage (Donald 1991).

Sutton has argued in his work on complementarity that a science of extended cognition should aim to develop taxonomies and typologies of external resources in use and of coupled agent-environment systems (Sutton 2010, p. 206; Sutton et al. 2010 – a task taken up by Heersmink 2014). He has used the idea of complementarity to argue for a dimensioned view of cognitive processes, according to which cognitive science should investigate the diverse and rich variety of ways in which cognition can involve the intricate coordination and interactive coupling of neural, bodily, and worldly elements (see also Sterelny 2010; Menary 2010). Such a dimensioned view is opened up once we recognise the variety of formats and radical heterogeneity of cognitive technologies humans can make use of in their thinking.[3] For example, we can distinguish between the form of the external resource – ecological, technological, or sociocultural (Wilson and Clark 2009). Different resources will have different influences on cognitive behaviour in part as a consequence of their material properties. Some external resources will become entrenched over time in an individual's action repertoire. Sterelny (2010) gives as an example a set of knives for a chef. Other external resources will have conventionalised uses that have been defined over long periods of cultural history and that are more easily transferable than individually entrenched resources (Stotz 2010).

We can distinguish between different kinds and strengths of coupling, which in turn will yield differences in degrees of the extension and distribution of cognition. Some interactive couplings will prove relatively fleeting, transient, and temporary; others may be more durable and persisting. Some couplings may occur only once in the life of an individual, never to be repeated; others may be reliable strategies for solving a problem that are regularly repeated by an individual or within a group (Theiner et al. 2010).

A alternative formulation of second-wave EM, found in the writings of Mark Rowlands and Richard Menary, focuses on cognitive integration and normatively regulated bodily manipulations of environmental structures, such as artefacts and tools. Cognitive processes are not located exclusively within the individual, because they are realised in part through bodily actions in which the agent engages their skill to actively transform or create structures in the environment (Rowlands 1999, 2010; Menary 2007). A familiar example is the rotation of zoids in Tetris to determine whether a particular zoid can be placed into a specific slot (Kirsh and Maglio 1994). The active manipulation of the zoid in this example accomplishes information processing that would otherwise have to be done through a computationally costly combination of mental imagery, working memory, and attention. The "epistemic" actions[4] that the agent performs do not solve the problem all by themselves; they must be coordinated with internal working memory and attentional processes. It is through the coordination and integration of the bodily manipulations of external structures (the rotation of the zoids) with internal working memory and attentional processes that the player solves the problem of where to place the zoid. The manipulation thesis claims that cognitive processes are extended when they involve the coordination and integration of embodied skilled engagements with external structures and internal neural processes. Extended cognition is found whenever we find this type of hybrid process that integrates internal neural processes with external processes of skilled bodily engagement with environmental structures (Menary 2010, p. 229).

Menary stresses that the bodily actions that the agent performs are not simply an alternative means of realising the information-processing functions that would otherwise be performed by the brain. The agent is enacting skills developed for the specific purpose of improving their performance in the game:

> The expert Tetris players. . . have developed motor programmes for manipulating the buttons that transform the zoids on the screen. As experts, their well-trained body schemas result in fluid and fluent actions, they do not need to consciously rehearse the next action.
>
> (Menary 2010, p. 575)

A great deal of human cognition depends on skills for creating, maintaining, and manipulating vehicles of public representations, such as writing, systems of mathematical notation, diagrams, and so on. Thus, consider the often-used example of using pen and paper to do multiplication (Rumelhart, McClelland and Hinton 1986). This cognitive shortcut requires making use of a notational system and the skill of writing. By combining our internal cognitive processes with cultural notational systems, cognitive agents are able to perform calculations more efficiently and economically than they would be able to were they to rely only on their own onboard capacities for arithmetical reasoning. Mastery of the notational system and the technique of writing are examples of what Menary calls "cognitive practices" (2007). Through socially scaffolded learning, we acquire capacities to produce public systems of representation, such as language and mathematics, and to exploit such systems for cognitive ends (see also Dutilh Novaes 2012). Menary (2007) has described the developmental process humans go through in learning such systems of public representation as one of "cognitive transformation." The development of such capacities is genuinely transformative of an individual's cognitive potential in that it allows for the integration of these public systems of representation with internal neural processes. One result of this integration is plastic changes in the brain that allow it to function in ways that are coordinated with externally located systems of representation.[5] Neural circuits "acquire new culturally specified functions, functions that have existed for only thousands, not millions of years" (Menary and Kirchhoff 2014, p. 5). These new functions are transformative of an individual's cognitive abilities in the sense that the agent is able to solve problems and engage in patterns of inference that would not be possible were it not for their integration of public systems of representation.

Menary argues that cognitive transformation is hard to make sense of from within the first wave of EM that stresses the coarse-grained functional similarity of inner and outer elements. External representational systems do not make a similar functional contribution to that of internal neural processes. They bring about cognitive transformation precisely because their integration yields a "different kind of functionality" that can be accomplished only through the skilled engagement and bodily manipulation of external, environmental structures. By placing the emphasis on functional similarity, the first wave risks failing to recognise how bodily manipulations alter the informational and physical structure of the cognitive niche, thereby transforming human cognitive capacities (Kirchhoff 2012; Menary and Kirchhoff 2014).

There is, however, a tension (though not necessarily an incompatibility) at the heart of the second wave between, on the one hand, stressing the transformation that cognitive integration brings about and, on the other hand, maintaining that internal and external elements make different but complementary functional contributions. The latter idea of complementarity can seem to imply that internal and external elements have relatively fixed and stable functional properties that remain the same across time. The latter claim is in tension with the transformation in cognitive capacities over the course of cognitive development that cognitive integration brings about. Such a transformation implies a more dynamic picture of the functional profile of the inner and outer according to which both are continuously and reciprocally influencing each other. We see such reciprocal influence particularly in sociocultural cases of extended cognition. Individuals through their activities collectively construct, maintain, and refine the social and technological systems that get taken up in cognition. At the same time, these social and technological systems support the development and exercise of cognitive practices. As Cash (2013) has nicely put it, "We are individually products, and collectively are producers, of these cognitive institutions, tools and practices" (2013, p. 64). The tension in the second wave between ideas of complementarity and transformation finds its resolution in the third wave of HEC, to which we now turn our attention.

1.4 Third-wave extended mind

The idea of a third wave in EM was first anticipated by Sutton (2010):

> If there is to be a distinct third wave of [EM], it might be a deterritorialized cognitive science which deals with the propagation of deformed and reformatted representations, and which dissolves individuals into peculiar loci of coordination and coalescence among multiple structured media . . . Without assuming distinct inner and outer realms of engrams and exograms, the natural and the artificial, each with its own proprietary characteristics, this third wave would analyze these boundaries as hard-won and fragile developmental and cultural achievements, always open to renegotiation.
>
> (p. 213)

Table 1.1 Key tenets in third-wave extended mind – expanded version

1. *Dynamic Singularity and No Fixed Properties*: some cognitive processes are constituted by causal networks with internal and external orbits comprising a singular cognitive system. In other words, cognitive processes have no fixed properties but comprise different orbits across different dimensions.
2. *Flexible and Open-Ended Boundaries*: the boundaries of mind are not fixed and stable but fragile and hard-won, and always open to negotiation. This tenet follows from tenets 1 and 3, as explained below.
3. *Distributed Assembly*: the assembly of cognitive systems is not always orchestrated by the individual agent but is sometimes distributed across a nexus of constraints, where some constraints are neural, some are bodily, and some are environmental.
4. *Diachronic Constitution*: cognition is intrinsically temporal and dynamical, unfolding over different scales of behaviour.

This quote specifies that the third wave of extended cognition is founded on several key claims. We presented these key claims or themes in the introduction to this book. Here we unpack them as they apply to third-wave extended mind (Table 1.1).

1.4.1 Dynamic singularities and no fixed properties

Third-wave EM denies that internal and external elements of extended cognitive systems have their own "proprietary characteristics" always and everywhere. It investigates instead the ways in which these elements are "deformed" and "reformatted" over different timescales given agent-environment interactions. It thus rejects what we will call the fixed-properties view of internal and external elements, which we have seen is at least implicit in the idea of complementarity (Kirchhoff 2012). In place of such a fixed-properties view, third-wave EM conceives of the properties of internal and external elements as continuously undergoing diachronic transformation. Cognitive processes should be viewed as dynamic and temporal trajectories of activity or equivalently as "extended dynamic singularities," a term we use in what follows, which in line with its usage in Hurley (1998). She uses this term to refer to a singularity in "the field of causal flows characterised through time by a tangle of multiple feedback loops of varying orbits." Such a dynamic singularity she says "is centred on the organism and moves around with it, but it does not have sharp boundaries" (Hurley 1998, p. 2).[6]

It ought to be unsurprising that Sutton's call for a third-wave formulation of EM echoes central and earlier points found in Hutchins's *Cognition in the Wild* (1995). Hutchins also emphasises that cognition is the propagation of activity across various media (where media must be taken to include bodily and neural hardware in addition to cultural elements). These assorted media are coordinated by a "lightly equipped human" working (sometimes) in groups, always embedded in cultural practices, where some of these dynamics become embodied in the individual cogniser over different temporal scales (Latour 1996, p. 62).

We should be careful not to misunderstand talk of processes of "embodiment" and "transformation." This should not be understood as if some *thing* moves across an environment-organism boundary. It will not do to think that the brain internalises cultural features by literally coming to encode cultural items after a process of internalisation and neuronal transformation. This would be to misinterpret how an agent comes to embody the regularities of its niche and how the agent is profoundly transformed by participating in cultural practices. According to Hutchins,

> [W]hat moves is not a thing, and the boundary across which movement takes place is a line that, if drawn too firmly, obscures our understanding of the nature of human cognition. Within this larger unit of analysis, what used to look like internalization now appears as a gradual propagation of organized functional properties across a set of malleable media.
> (1995, p. 312)

"Internalisation" is thus not a matter of moving the forms of knowledge enacted in cultural practice inside of the head of an individual. It is instead to be understood in terms of a reconfiguring of the individual through the network of practices in which the individual participates. Following in the footsteps of Hutchins, third-wave EM takes the external environment to transform what agents can do cognitively, both in the synchronic here and now and over diachronic timescales (though in the end these timescales cannot be separated) (Menary 2007; Vygotsky 1978; Wertsch 1985). What is transformed is not merely the internal cognitive functions of the individual's brain so that once the individual's brain is transformed, the cultural practices can then be thrown away, having accomplished all their work. It is only through ongoing participation in culturally organised activities with other people that the individual agent can be said to undergo cognitive transformation (Malafouris 2010; Sutton et al. 2010; Hutchins 2011). Properly understanding internalisation requires looking beyond the system comprised of the individual and the external resource or element. It requires looking at the individual situated in a wider set of already existing and ongoing cultural practices.

Thus, internal systems do not have their own proprietary characteristics that become transformed through enculturation. Internal cognitive functions undergo transformation only through an individual's continuing participation in cultural practices. Nor do external resources have their own fixed properties. Consider in this light the distinction between engrams (biological memories) and exograms (culturally encoded public representations). It would be a mistake to interpret this distinction as implying that exograms are memory records located outside the brain. This would be to overlook the place that such records occupy within cultural systems of representations (Hutchins 2008, 2011). Any external encoding of information only counts as a representation with a determinate meaning because of cultural norms that govern the usage of these representations and the embodied activities of individuals that enact those norms (see also Hutto and Myin 2017). Exograms do not have some fixed pre-given format. They should instead be thought of as continuously reformatting through the meshing of embodied actions, physical materials, and cultural norms of usage (Malafouris 2010).

1.4.2 The flexible and negotiable boundaries of the mind

Hurley describes the dynamic singularity as "organism-centred" (1998, p. 2). The external and internal feedback loops are centred on a single individual. At the same time, she allows for the individual agent to have no sharp boundaries, a claim that Clark (2008) echoes in his argument that cognitive processes are organism centred but not organism bound. Without abandoning the idea that internal and external cycles of feedback all feed into a single individual, the third wave rejects what Clark (2008) has labelled "the hypothesis of organism-centred cognition." Clark argues that the processes by which cognitive systems are softly assembled is centred on the individual agent or organism. It is the individual that in the end determines whether to solve a problem by relying exclusively on internal onboard cognitive machinery or to instead exploit structures in the environment. Thus, Clark writes:

> Human cognitive processing (sometimes) literally extends into the environment surrounding the organism. But the organism (and within the organism, the brain/CNS) remains the core and currently the most active element. Cognition is organism-centred even when it is not organism-bound.
>
> (2008, p. 139)

The third wave calls this assumption into question, proposing instead to dissolve the cognitive agent into a "loci of coordination and coalescence among multiple structured media" (Sutton 2010, p. 213; see also Hutchins 2011). The boundaries separating the individual from their environment and from the collectives in which the individual participates are "hard won and fragile developmental and cultural achievements" (Sutton 2010, p. 213). While an organism-centred view of cognition takes cognition to always be centred on an individual cognitive agent, the third wave leaves this as an open issue to be settled on a case-by-case basis. Hutchins expresses this point well when he writes that "Some systems have a clear centre, while other systems have multiple centres or no centre at all" (2011, p. 5).

The view of cognition as organism centred can be cast as implying the fixed-properties view of the components of cognitive systems that the third wave rejects. Internal psychological processes do the work of recruiting external resources for use in cognition. All of the active work in recruitment is done by the individual organism, rendering anything on the side of the environment causally inert and static so far as the assembly process is concerned (Hutchins 2011; Kirchhoff 2015). It is not clear, however, that this unidirectional model of causal influence is the right way to think about cognitive assembly. As Menary stresses, plastic brain processes undergo profound changes in their functioning through the process of cognitive integration. Menary's manipulation thesis requires us to pay close attention to the sculpting of neural processes through development to work in collaboration with environmental structures (2015). The fixed-properties view is the result of thinking about the couplings through which extended cognitive systems are assembled synchronically over a short timescale of action in the here and now. Once we shift to thinking of extended cognition diachronically, as talk of cognitive transformation encourages us to do, we see that the fixed-properties view cannot be right. It is this diachronic perspective on extended cognition that stands behind the third wave's rejection of the hypothesis of organism-centred cognition.

1.4.3 Distributed cognitive assembly

The third wave follows Hutchins (2008, 2011) in taking the patterned activities that people enact by taking part in cultural practices to play a role in cognitive assembly. The individual is not the loci of control and coordination in cognitive assembly (see also Kirchhoff and Newsome 2012). Cognitive assembly is a self-organising process, and control and coordination are distributed over and propagated through the media taken up in cultural patterns of activity. A self-organising process is one in which a set of components that make up a system enter into non-linear interactions according to local rules, without the intervention of any global executive control process (Kelso 1995). Out of the non-linear interactions among the system's components, order and structure arises and is maintained in the system as a whole. The constraints (the local rules) that govern the interactions between the components (internal and external) of extended cognitive systems need not all arise from within the biological organism. The third wave argues that some of the constraints may originate in social and cultural practices, in "the things people do in interaction with one another" (Hutchins 2011, p. 4). That is, some of the constraints that govern self-organisation of the elements that make up cognitive systems arise from neural mechanisms; others from material tools, artefacts, and technologies; and still others from social practices and conventions. Cognitive systems configure themselves based on the interaction among these constraints and mechanisms for constraint satisfaction (Hutchins 2014, p. 13).

In this regard, humans are more like ants or termites than spiders: usually we need not spin our own cognitive webs because they have already been spun in the cumulative activity of previous generations (Sterelny 2012). Thus, in Elizabethan theatre practices, actors were able to take part in up to six different plays per week (Tribble 2005). Actors managed to remember their parts in part through relying on cues from a cleverly structured niche. Relevant to the present discussion, however, is the way in which novice actors were trained by more expert actors. The process of cognitive assembly in the novice actors was arguably distributed over the novice, the expert, and the structures in the theatre they were relying on in their performance. Even in fully trained actors, it makes sense to think of their recruitment of structures in the theatre as taking place in the wider context of the sociocultural practices and technologies of the Elizabethan theatre world.

The organism-centred view of extended cognition that Clark defends implies that cognitive processes typically unfold inside of the individual and occasionally go out into the world. Clark suggests (drawing on research by Gray and colleagues (Gray and Fu 2004)) that the brain makes a cost-benefit calculation to determine whether the most efficient strategy is to rely on one's onboard machinery or whether a more efficient solution relies on some combination of biological wetware plus situated action. The output of this cost-benefit analysis will determine whether a given cognitive task is performed by relying exclusively on the brain, or if it is performed partly in the world through the agent's situated actions.

The third wave suggests a different starting point. It conceives of the individual agent taken as a whole: the entire embodied organism, including its brain and central nervous system, is conceived of as plastic – that is, open to continuous revision and configuration. Moreover, the individual agent is embedded in a plastic network of relations and processes, whose dynamics break across organism and environment. This means that any alleged cognitive division between organism and environment becomes much more vague, ephemeral, and fragile. In other words, cognitive systems – both their unique properties and distinctive boundaries – are not biologically fixed but rather develop and unfold with a bio-cultural system subject to constant transformation (Kirchhoff and Meyer 2017; Malafouris 2010). Crucially, third-wave HEC is thus committed to the view that cognitive systems are first and foremost nested, multilayered systems, whose boundaries continuously shift over time to include elements spatially located outside of the individual in its niche (see also Stotz 2010).

1.4.4 Diachronic constitution

Most philosophers agree that causation is different from constitution. We have presented third-wave EM as the position that the constitutive basis for extended cognitive processes is diachronic, as opposed to synchronic. By this we mean that extended cognitive processes are composed of elements unfolding over multiple different timescales. This follows from denying the fixed-properties view and the organism-centred view of cognitive assembly. The fixed-properties view was rejected on the grounds that the elements of a cognitive system undergo continuous deformation and reformatting through agent-environment interaction. Cognition is characterised in terms of a dynamic and temporal trajectory of activity that propagates over various media: some neural or bodily, others involving other people and the resources provided by an environment shaped by our cultural activities and patterns of practice. The organism-centred view of cognition was rejected on the grounds that it is through the temporally extended process of bringing bodily actions and artefacts into coordination based on constraints that arise in cultural practice that cognitive processes extend.

Much of the discussion of the causal-constitution distinction as it applied to cognitive extension has proceeded as if constitution were a synchronic relation between a system and its constituent parts. Based on such a synchronic understanding, all of the constituent elements ($E_1 \ldots E_n$) of a system *S* must be wholly present at each instant that *S* exists. Such a synchronic conception of constitution is, however, ill-suited to understanding the relation that holds between a cognitive process and the subprocesses of which it is composed. Each subprocess unfolds continuously over time, exhibiting its own rate of change, rhythm, and duration, and will be influencing and be influenced by the other subprocesses of which *S* is composed (Van Gelder and Port 1995). The constituent subprocesses may partially overlap in time, but in order to contribute to the realisation of a system *S*, it is not necessary for their existence to entirely overlap with that of *S*. Crucially, the subprocesses that make up the system do so over temporally extended intervals, not at discrete instants in a stepwise and linear manner. The notion of constitution that is needed for understanding a temporally complex phenomenon like cognitive extension is a diachronic notion of constitution (Kirchhoff 2015, 2015a, 2015c).

Clark's restriction of extended cognition to short timescales screens off the fact that even in the here and now, history and culture are always embedded and carried along in the practices and artefacts that individuals are engaging with (Menary 2007; Sutton 2008, 2010; see also Haugeland 2002; Lave and Wegner 1991). Clark casts history and culture in the role of background conditions that set the stage for the brain to do the work of softly assembling cognitive systems. In doing so, he seems to assume that all of the work of cultural practices in constraining, coordinating, and self-organising action has been fully internalised. What is learned from others through training in social practices is internalised in the form of internal representations and then gets to do its work through the brain. The cultural transmission of knowledge and practices is understood as transmitting information among individuals. Once the information has circulated in the right way among individuals such that the practices are learned and adequately internalised, there is no longer any work left for the cultural practices to do.

Once we recognise that the notion of constitution required for understanding extended cognition is diachronic, there is no need to screen off cultural practices and other people in this way. Internalisation is only ever partial and provisional (Latour 1996). It is partial because some of the constraints that are satisfied in the self-organisation of an extended cognitive system arise in practice. It is provisional because cultural practices are dynamic and should not be reduced to a collection of lifeless artefacts. We can allow that the individual agent does indeed establish and maintain dynamic coordination among the multiple elements of which a cognitive system is composed: some internal, some external to the individual. However, the individual does so based on constraints that are imposed on an individual by cultural patterns of activity. The individual is thus "the loci of coordination," but the control of its activities is decentralised and distributed among "multiple structured media and practices" (Sutton 2010, p. 213). It is the individual as embedded in a network of practices that is reconfigured in their coupling with the environment.

Cultural practices guide what an individual attends to and thus how they perceive the world as well as what they are ready to do – their patterns of action readiness. When a chimp like Sheba acquires competence with a system of symbols (e.g., number symbols), it is obviously true that she must acquire certain internal capacities such as capacities for recognising the symbols and understanding and interpreting them (Boysen et al. 1996; Hutchins 2008). However, what she does with the symbols as she interacts with her keepers is not wholly down to her but is in addition due to the preexisting constraints and norms that govern the symbol-using practice into which she is initiated by her keepers. The extended cognitive system of Sheba and the external number symbol system self-organises and assembles itself in part based on constraints that come from the symbol-using practice. The constraints originating in the practice have what Haugeland (2002) nicely termed a "narrative gravity." When Sheba uses the symbols in line with the norms of the practice, or close enough, she is acting within the orbits defined and delineated by the norms of the practice. If her behaviour wanders too far astray from the orbits defined by the practice, she is corrected and "pulled back in" (Haugeland 2002, p. 32) by her keepers.

Clark's treatment of wider cultural practices as a background condition for the assembly process that takes place in the here and now is unjustified. Cultural practices should instead be included within the processes unfolding here and now because individual agents in many cases of extended cognition are embedded in wider networks of cultural practices. The cognitive task is performed within an interval of time in the here and now. Insofar as the individual is embedded in a set of practices, practices are among the causal factors that influence how the elements of a cognitive system organise within this interval of time. Thus, some of the active dynamic processes involved in the process of softly assembling a cognitive system are located beyond the individual agent, in the wider practices in which the individual is nested.

1.5 Summary

Our discussion in this chapter has been restricted to unconscious cognitive processes. EM has been mostly articulated in terms of dispositional belief-like states such as the example of Otto and his notebook and by appeal to unconscious perceptual states – more specifically, those that accompany epistemic actions, such as in the Tetris example discussed above. When consciousness has been discussed, theorists otherwise sympathetic to the extended mind have tended to deny that phenomenal consciousness is extended, partially constituted by engagement with elements located in the environment. For instance, Clark has argued that consciousness qualifies as a type of cognitive process that can never extend into the world (see also Wheeler 2015a, b). He has argued (quite consistently with his vigorous support for the thesis of the extended cognition) that the "material apparatus" that supports conscious mental life "can still quite reasonably be thought to be wholly internal" (Clark 2012, p. 755). In the next chapter, we review the arguments that have been made for extended consciousness. These arguments come from philosophers sympathetic to sensorimotor enactivism.

Notes

1 We use the terms "constitution" and "realisation" interchangeably.
2 For a different example of the application of the parity principle, recall the by now familiar case of neurobiologically impaired Otto and his notebook. Otto suffers

from a mild case of Alzheimer's disease and has developed the habit of writing down useful information in a notebook akin to how others might be said to retain information in long-term memory. According to Clark and Chalmers, because the dispositional information in Otto's notebook is functionally poised to guide action in a functionally similar way as non-occurrent beliefs in biological memory, the information in Otto's notebook should be considered cognitive belief-like states. Importantly, the parity principle by itself is not meant as an argument for EC but rather to encourage us to look at various cases of cognitive extension "behind a veil of metabolic ignorance" (Clark 2011, p. 449).

3 The idea of such a dimension is by no means exclusive to the second wave. Clark and Chalmers (1998) attempted to specify various criteria for treating external objects as parts of a cognitive system. They suggested that the external element should, for instance, be reliably accessed and more or less automatically endorsed. These conditions have come to be referred to as the "glue and trust" conditions. Notice, however, that they come in degrees and thus admit a spectrum of cases that can vary across the dimensions of the degree to which the resource is trusted by an individual and is accessible to them.

4 Kirsh and Maglio make a distinction between "pragmatic" and "epistemic" actions, where the latter are "actions performed to uncover information that is hidden or hard to compute mentally" (Kirsh and Maglio 1994, p. 513).

5 One example of evidence for such a claim comes from Dehaene and colleagues' work on neuronal recycling, in which they show that "cultural acquisitions are only possible . . . by reconverting pre-existing cerebral predispositions for another use" (Dehaene 2005, p. 134). Thus, Dehaene and colleagues have shown that capacities for exact mathematical reasoning depend on culturally acquired language-formatted representations (Dehaene et al. 1999). In development, a subset of the brain's cortical representations undergoes a transformation from a primarily non-linguistic to a linguistic mode of functioning. It is this transformation consisting of neural processes coordinating to culturally acquired forms of representation that makes possible exact mathematical reasoning.

6 We adopt Hurley's notion of a 'dynamic singularity' because it aligns with thinking of both the self-organisation and nature of mental processes in temporal and diachronic terms, on the one hand, and with the claim that such processes are realised in an extended network of processes, on the other. We discuss the question of whether the self-assembly of dynamic singularities needs to be cast as centred on the organism in Chapter 5 and Chapter 6.

2 From extended mind to extended consciousness?

2.1 Introduction

The thesis of extended consciousness claims that the biological machinery that realises conscious experience can sometimes include a mixture of neural, bodily, and environmental elements. It is a thesis about what is and is not part of the material substrate of a creature's qualitative mental life (Block 2005, p. 264). The notion of consciousness at stake is phenomenal consciousness. It is "the elusive 'what-it-is-likeness' that seems to characterise a subject's experience of a certain kind of redness, of a certain voice, or of a pain in her stomach" (Clark 2009, pp. 963–964). Phenomenal consciousness is the topic of much puzzlement and wonder among scientists and philosophers. It is an undeniable feature of human and non-human animal reality, and yet how this can be has struck many as mysterious. Why is it that when an animal has a physical and biological organisation like ours it enjoys conscious experiences of its world? Could there not exist a creature that was a duplicate of us with respect to its physical and functional organisation, that consequently behaved just as you and I do, but that experienced nothing whatsoever of its existence?

Philosophers disagree about what to say about such imaginary beings. In what follows, we set aside these difficult and unresolved questions. We take for granted that it can be made scientifically intelligible why subjects undergo the particular phenomenal experiences they do, or any experience at all. We do so for reasons explained in our introduction. We must make this assumption to get to the questions that interest us in this book about where consciousness is realised when it is realised.

Defenders of extended consciousness ask why we should think the boundary of skin and skull is somehow privileged so that it is only processes taking place within this boundary that can realise conscious experiences? As Hurley put it, "Why should dynamics distributed within a pre-specified boundary be capable of explaining qualities, while those beyond are in principle ineligible?" (2010, p. 112). In a similar vein, Noë (2009) has argued that the biological substrate of consciousness is the whole organism in an environment. Consciousness, he suggests, requires "the joint operation of the brain, body and world"; it is "an achievement of the whole animal in its environmental context" (2009, p. 10). Evan Thompson has suggested that consciousness is no more generated by processes inside of the head than flight is generated in the wings of a bird. Neural processes are necessary for consciousness, just as wings are necessary for a bird to take flight. But flying is an activity of the whole bird in its environment. Wings help the bird to fly but cannot do all the work on their own. Similarly, it is only if neural processes are coupled in the right way to a body in the world that we get the kinds of experiences we typically enjoy. A brain that was not embodied and embedded in the natural world in the way we are might well be able to support some kind of experience. But it could probably not support the kind of phenomenologically rich, stable, and detailed world-presenting experience that we typically enjoy.[1]

Our aim in this chapter will *not* be to present the details of the third-wave arguments for extended consciousness. This will be done later in the book (Chapters 3 to 6). Here our agenda is threefold. First, we aim to provide an overview of the debate surrounding extended consciousness as it has played out in the literature so far. We do so in part for readers that are unfamiliar with the details of this debate. Second, we showcase some of the reasons why almost all defenders of EM have denied the existence of extended consciousness (Chalmers 2008, forthcoming; Clark 2009, 2012; Wheeler 2015a, 2015b). Arguments for extended consciousness have tended to come from defenders of sensorimotor enactivism, not from defenders of EM (e.g. Hurley 1998; Noë 2004, 2009; Ward 2012). According to sensorimotor enactivism, conscious experience of the world is causally enabled by cycles of perception and action in which agents dynamically couple with their environment. Our third aim is to provide a worked-out account of how sensorimotor enactivism might be taken to imply extended consciousness. In doing so, we will be led to an argument for extended consciousness. This argument will set the stage for the challenges we take up in the remainder of the book.

2.2 Sensorimotor enactivism and extended consciousness

Sensorimotor enactivism takes the qualities of phenomenal experience to be fixed by the dynamics of agent-environment interaction (Hurley 2010). It is first of all a theory that seeks to explain *what* subjects experience as partly externally realised. We will thus follow Hurley (2010) in characterising sensorimotor enactivism as a variety of externalism about the qualities of experience. Consequently, it contrasts with internalist theories that take the qualities of experience to be fixed by properties internal to individuals.

Internalists hold that two individuals could inhabit radically different environments – one could be a brain in a vat, for instance – but still undergo qualitatively indistinguishable experiences. Externalists about the qualities of experience hold by contrast that qualities of at least some of our experiences would vary across individuals inhabiting different environments.

Block (2005) imagines a brain that is a duplicate of my own springing into existence and floating in space. Would not such a duplicate have experiences just like mine? Externalists are committed to answering this question in the negative. If the brain is not hooked up and plugged into an environment just like mine, because it is floating disembodied in space, it could not have experiences just like mine (Noë 2009). According to sensorimotor enactivists, this is because the qualities of experience are realised by the dynamics of sensorimotor interaction with the environment. The dynamics of the agent-environment coupling is tight, such that the parameters of one system (the agent) are the variables of the other (the environment) and vice versa (Beer 2003). Thus, it will not always be possible to "unplug" the internal neural factors that bring about experience from the environment and "replug" them into a different environment without this replugging changing the functioning of the internal neural factors (see Box 2.1 for details on coupling).[2]

So-called what-explanations – explanation of what subjects experience – have implications for the explanations that are given of *how* phenomenal experiences are realised (Ward 2012). In particular, sensorimotor enactivists hold that it is neural activity as "embedded in an animal's larger action and interaction with the world around it" that matters for the realisation of some types of phenomenal experience people typically have (Noë 2009, p. 47). What the brain does in all of this is facilitate the cycles of perceptual engagement with the world, just as the bird's wings facilitate flight.

In the next subsection, we will survey how sensorimotor enactivists account for the qualities of phenomenal experience in terms of the dynamics of sensorimotor interaction with the environment. The remainder of the chapter then aims to clarify how this externalist theory of what we experience can be used to argue for an externalist variety of how-explanations – the thesis of extended consciousness.

2.2.1 The no-magical-membrane argument

Sensorimotor enactivists argue for externalism about the qualities of experience first by pointing out that there is no good reason to believe in a sort of magical membrane such that anything behind this membrane has the power to generate phenomenal experience and anything outside of it does not. We know that phenomenal experience does not arise based on activity in small collections of cells. It likely depends on large-scale dynamical activity that

Box 2.1 Agent-environment couplings

Beer (1995) provides an early dynamical model of two coupled dynamical systems, which he associated with an agent and its embedded environment. Beer brought to the forefront the idea that the brain can be modelled as a system within the body, and a coupled brain-body system can be modelled as a jointly coupled system embedded within a larger system, the environment (Figure 2.1).

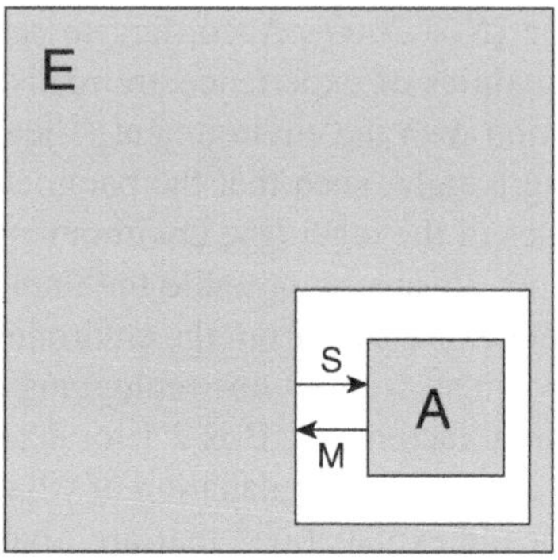

Figure 2.1 An agent and its environment as coupled dynamical systems

Source: adapted from Beer 1995, p. 182

Under simplifying conditions, Beer (1995) modelled an agent and its environment as two continuous-time random dynamical systems, *A* and *E*. Formally, the coupling between *A* and *E* can be mathematically represented as follows:

$$x_A = A(x_A; S(x_E); u'_A);$$

$$x_E = E(x_E; M(x_A); u'_E)$$

S represents a sensory function from environmental variables to parameters of the agent, and *M* represents a motor function from agent state variables to parameters of the environment. The notation $S(x_E)$ is associated with the sensory inputs of an agent and $M(x_A)$ with motor outputs. Input arguments u'_A and u'_E correspond to any residual parameters of *A* and *S* respectively that are not taken to play a role in the coupling.

Feedback plays a crucial role. As Beer emphasises, "Any action that an agent takes affects its environment in some way through *M*, which in turn affects the agent itself through the feedback it receives from its environment via *S*. Thus, each of these two dynamical systems is continuously deforming the flow of the other" (1995, p. 182). From the kind of coupling highlighted between *A* and *S*, Beer goes on to claim that it is possible to view the two coupled dynamical systems assembling a larger system *U* whose state variables are the union "of the state variables of *A* and *S*" (1995, p. 183).

unfold over time and that spread across the brain as whole. The late Francisco Varela refers to this as the "brainweb" (Varela et al. 2001). This large-scale neuronal activity involves complex recurrent and re-entrant dynamics such that it is difficult, if not impossible, to say where inputs end and outputs begin (Thompson 2007). Complex looping dynamics are, however, not confined to the brain. The same kind of looping dynamics unfold when a perceiver engages in sensorimotor interaction moving their eyes, head, and whole body as they explore the environment. As you move your body, what you sense of your environment varies. Embodied activity changes the patterns of energetic stimulation detected by the organism's perceptual systems, and this energetic stimulation in turn influences your embodied activity. Sensory and motor channels are thus continually and reciprocally influencing each other in ways that depend on agent-environment interaction. They stand in a relation of reciprocal co-determination.

Now using these points, we can make a parity-style move to argue for what-quality externalism because what we have just shown is that feedback loops that occur inside of the brain also occur across the agent-environment boundary given the coupling of sensory and active states. If recurrent, re-entrant dynamics can fix the qualities of experience when they unfold inside the brain, why can they not do so when they loop outside the brain through cycles of sensing and acting? Hurley writes that

> We're familiar with the idea that explanatory processes can be distributed across desperate areas within the brain instead of being localised. But no magical membrane contains distributed processing; brains are in continuous causal interaction with their bodies and their environments. Why should dynamics distributed within a pre-specified boundary be capable of explaining qualities, while those beyond in are in principle ineligible?
>
> (2010, p. 116)

We call this the no-magical-membrane argument. Like the parity argument in first-wave extended mind, this argument has the objective of undermining intuitions that the biological boundary separating the individual from the rest of the environment is somehow special when it comes to generating consciousness. But unlike the parity argument, which focuses on non-conscious cognitive states such as dispositional beliefs, the no-magical-membrane argument targets the qualities of experience.

Hurley presents the argument in the course of outlining what she calls the "autonomy metaintuition for phenomenal qualities." This is the intuition that there is an explanatory gap separating the phenomenal qualities of experience from functional and physical properties. Hurley's point is that if one has little or no idea how functional and physical features could generate phenomenal experience, this puts one in a weak position when it comes to ruling out externalist explanations, such as those of sensorimotor enactivism (Hurley 2010, p. 104). In other words, if one has little or no idea about how consciousness arises in the first place, then one has no justified reason to privilege the brain as the seat of consciousness at the expense of the rest of the body and world.

If one believes in an explanatory gap, however, one is unlikely to be moved by considerations of this sort. One might well respond that both internalist- and externalist-style explanations are in an equally hopeless position when it comes to explaining the qualities of experience. Both leave us equally in the dark when it comes to explaining why experiences have the qualities they do or any qualities whatsoever. If the no-magical-membrane argument is to have any force against such worries, it needs to be combined with an externalist explanation of what subjects phenomenally experience. In the next subsection, we provide such an explanation.

2.2.2 The variable-neural-correlates argument

Sensorimotor enactivists have attempted to provide such explanations by appealing to two kinds of cases, both involving neural plasticity. We will refer to these as the variable-neural-correlates arguments. In the first, we are presented with examples in which there is variability in neural processing while the subject nevertheless enjoys experiences of the same phenomenal type. One such example is the rewiring experiments that Sur and colleagues carried out on newborn ferrets (Sur et al. 1999). Cells in the eyes that would normally grow connections to visual areas of the ferrets' brains were rerouted to areas that would typically deal in auditory signals. The result of this intervention was that the ferrets grew up seeing with parts of the brain that would normally be involved in hearing. This is a case of variable neural activity but sameness in the character of experience – seeing and not hearing, the ferrets do not hear with their eyes because of the rerouting of visual information to auditory cortex.

The second type of case sensorimotor enactivists present has the opposite profile. We are given examples in which subjects undergo a change in qualitative experience while their neural processes remain constant. Sensory substitution is one such example. Sensory-substitution devices are systems that, following a period of training, provide the visually impaired with a quasi-visual mode of access to the world analogous to vision (Bach-y-Rita and Kercel 2003; Kiverstein et al. 2015). They work by converting images captured by a camera into electrical and vibratory stimulation or sound frequencies that are then delivered to a normally functioning sense like touch or hearing. Through training, the perceiver discovers patterns in the sensory stimulation that provide them with a mode of access to the world analogous to vision.

Sensory substitution is given as an example in which neural processing remains constant but experiences change in their character. It is argued that neural processing remains constant because information from the camera converted into tactile stimulation continues to be processed in somatosensory cortex.[3] Once the user has mastered the device, however, they cease to notice the tactile stimulation on the surface of their skin and begin to gain a sort of quasi-visual access to the distal objects and the properties around them at a distance from their body. It does not matter for our purposes whether we characterise this change in experience as a change from touch to seeing. What is important to the argument is that the subject undergoes some sort of change in experience while their neural processing remains unchanged. That is to say, neural processing continues to take place in somatosensory cortex.

In the first set of cases, experience remains constant while neural processing varies. In the second set of cases, experience varies and changes in character while neural processing remains constant. The question is then, how can we account for sameness or difference in experience in these examples? Such an explanation would potentially help us understand how to put experiences into equivalence classes. That is to say, it would help us to provide a what-explanation of the qualities of experience, of what makes it the case that phenomenal experiences are of the same or different types.

The point of these examples is that they are supposed to put pressure on the idea that the explanation could be given purely and exclusively in terms of neural processes. They are examples in which similarity or difference at the level of phenomenal experience cross-cuts similarity or difference at the neural level. Either we are given examples in which neural processes are the same but phenomenal experiences are different, or vice versa.

This argumentative strategy opens up an obvious line of attack from internalists. It might be objected that the first set of cases do not provide genuine examples of differences in neural processing. Thus, Aizawa has objected to Noë's treatment of the ferret examples along these lines by arguing that auditory cortex in the rewired ferrets takes on the structural properties of visual cortex (Aizawa 2010). Thus, we do not have an example of sameness in experience but difference in neural processing. Auditory cortex essentially becomes visual cortex in the rewired ferrets. Wheeler (2015a) has argued along similar lines that sensory substitution does not provide us with a case in which neural processing is constant but experiences change. Instead, he suggests that sensory substitution may induce "a fundamental change in the mathematical structure of the neural activation patterns in somatosensory cortex" (Wheeler 2015a, p. 169). Finally, it might be questioned whether in perceptual adaptation veridical experiences are really restored; it could instead be the case that motor behaviour adapts over time just like it does when you brush your hair using a mirror (Prinz 2008). Is experience really the same before and after donning the goggles?

Unsurprisingly, then, internalists have not been convinced by the variable-neural-correlates argument. Fortunately, we do not need to defend the argument against its many detractors. Our interest in this argument is instead in how it is used to motivate an externalist account of phenomenal experience. We concede that it fails as a refutation of internalism. Still we can raise the question that the variable-neural-correlates argument was introduced to motivate and assess the plausibility of an externalist answer. The plausibility of the externalist explanation of the qualities of experience does not hang on the success of the variable-neural-correlates argument.

Recall that the argument was introduced to provide us with real-world cases (as opposed to philosophical thought experiments) in which experiences either change or keep a constant phenomenal character. We can still ask, what makes it the case that an experience belongs to one phenomenal type rather than another? What makes the difference between seeing and touching in the case of sensory substitution, for instance? What makes it the case that experience changes in perceptual adaptation (if it indeed changes) from being illusory to being veridical? The answer that the sensorimotor enactivist proposes, as we shall see in detail in the next section, is that mastery of sensorimotor contingencies (SMCs) explains why experiences belong to one phenomenal type rather than another.

2.2.3 An externalist account of the qualities of experience

SMCs are regularities that relate sensory input to movement. Thus, we can make a distinction between movement- and object-dependent SMCs. Mastery of SMCs is marked by a perceiver's a high degree of familiarity with "not only the sensory effects of movement but also the sensory effects produced by environmental changes" (Noë 2004, pp. 64–65). Movement-dependent SMCs reflect the ways in which movements of the body affect sensory perception. Object-dependent SMCs relate to the movement of the object and how the object's movement changes a person's sensory perception.

There is an ambiguity in how to understand the term "sensory effects" in this characterisation of SMCs. It can be understood either proximally in terms of energy patterns impacting the sensory organs. Or, such sensory effects of movement can be understood distally in relation to their environmental causes. In ecological psychology, for instance, the ambient array that surrounds the perceiver and that the perceiver moves through provides information for the perceiver because it stands in systematic relations of covariation with surrounding surfaces that can be used to "pick up" on environmental affordances (Gibson 1966, 1979). If we understand SMCs in terms of the ecological ambient array, the systematic relations between sensory effects and movement need to be understood distally. They pertain to changes in the ambient array brought about through movement.

Such an ecological reading of SMCs is in line with the extensive use that Nöe (2004) makes of Gibson. Ecological readings of SMCs entail externalism, while proximal readings of SMCs are at best neutral on whether the explanation of the qualities of experience in terms of SMCs involves the environment. We will henceforth understand SMCs in ecological terms.[4] The possibility of giving an internalist reading of SMCs is something we return to later in this chapter and elsewhere in the book (Chapter 5).

According to sensorimotor enactivism, mastery of SMCs determines the phenomenal character of an experience. In perceptual adaptation, for instance, the perceiver gains a familiarity with SMCs that have changed as a result of wearing inverting goggles. Once they have familiarised themselves with and acquired mastery over how movement induces systematic changes in the ambient optic array as mediated by the goggles, their experience is predicted to adapt. It changes because they are once again able to track one and the same surface property (such as colour in the case of colour-inverting goggles) as its shows up under varying viewing conditions. According to sensorimotor enactivism, it is these types of tracking abilities that determine the phenomenal quality of experience. Insofar as they depend on familiarity with SMCs, it is SMCs that explain what phenomenal type an experience belongs to.

We see a similar line of argument in Noë's treatment of a perceptual phenomenon that he labels "presence in absence." The classic example is seeing the whole bulgy tomato even though the back of the tomato is hidden from view. I do not just see the parts of the tomato that are present to me from the point of view I currently occupy. I see the presence of the whole tomato, including its hidden aspects: the tomato looks to have a back currently hidden from view. This is how the tomato appears. I do not infer or judge that the tomato has a backside. I have a visual sense of its presence. Noë calls this visual sense "perceptual presence."

The concept of perceptual presence applies not only to the occluded parts of objects but more generally across the board. Everything a subject perceives is perceived from a point of view. Thus, the colour of an object is presented under certain viewing conditions in which there are variations in brightness, reflectance from other objects, and so on. Subjects nevertheless typically see the actual colour as present, although strictly speaking, these variations in viewing conditions make it the case that the actual colour is absent. When holding a coin between your fingers, you see the circular shape of the coin despite it being presented to you from an angle. The actual shape of the coin is present in absence.

As a final example, consider the visual field as a whole. We have a sense of seeing the environment as present in full detail and in high resolution. Perceivers can, however, fail to detect the presence of large changes when the change is accompanied by a distracting transient such as a flash. Perceivers can likewise fail to detect the presence of a stimulus such a man running across a basketball court dressed in a gorilla suit when they are not expecting it or when their attention is occupied by another task. We have an impression of seeing colour uniformly distributed across the visual field all the way from the centre out to the periphery. Yet outside the foveal region, the retina is near colour-blind, and our powers of discrimination are severely limited.

Noë has argued that the explanation of perceptual presence again lies in skills for tracking and locking onto objects and their properties that have their basis in mastery of SMCs. I do not represent all of the detail in the visual field in detail like in a picture. Instead, I have a sense of the detail as present out there in the environment, available to me to explore. The detail is perceptually present to me because of my mastery of SMCs. I understand that I can gain access to what is out there in the environment by moving my eyes or head.

Phenomenal experiences can thus be assigned to equivalence classes based on the objects and qualities perceptually present in experience. Perceptual presence is then explained in terms of practical mastery of SMCs. It is embodied skill or practical understanding that allows a perceiver to bring "the world into focus," "to lock onto it" (Noë 2012: p. 24). Perceptual consciousness is "transactional," Noë claims (borrowing a term from Dewey) (Noë 2012, p. 22). It involves a two-way relation to the world that depends on both one's current sensory relation to an object (e.g., the facing side of the tomato, the coin as presented from an angle) and one's sensorimotor skills (one's mastery of SMCs). This is an externalist explanation of the qualities of phenomenal experience, because both the sensory relation to an object and the operation of sensorimotor skills involve the perceiving animal's active relation to its environment. Noë writes that

> seeing is not something that happens to us. It is not something that happens in our brains. It is something we do. It is an activity of exploring the world making use of our practical familiarity with the ways in which our own movement drives and modulates our sensory encounter with the world.
>
> (2009, p. 60)

We set to one side for now the evaluation of this as an account of the qualities of experience. Our aim in this chapter is not to defend sensorimotor enactivism as an externalist account of the qualities of experience. Our aim is instead to show how such an account of the qualities of experience can be used to provide support for extended consciousness. Hurley (2010) notes that externalism about the qualities of experience does not obviously entail externalism about the vehicles of conscious experience. She suggests that it could consistently be combined with an internalist account of how sensorimotor skills are realised. Towards the end of the chapter and elsewhere in the book, we will discuss positions that aim for such a middle-ground view (see also Seth (2014) on perceptual presence, discussed in Chapter 5). Nonetheless, we think there is a line of argument from sensorimotor enactivism to extended consciousness. In the next section, we will review this argument and attempts at refuting it.

2.3 The DEUTS argument for extended consciousness

In 2009, Clark published a landmark paper in *Mind* asking whether the arguments for the extended mind might generalise to phenomenal experience. Recall that the thesis of extended consciousness is a thesis about how phenomenal experiences are realised. Phenomenal experience can be argued to depend on processes that extend into the environment if some phenomenal experiences are physically realised in processes that extend beyond the brain, into the body, and out into the environment. The internalism that interests us in this section is the mainstream view that phenomenal experiences are fully and wholly realised by neural processes inside the heads of conscious creatures. Such a view is admittedly much more intuitive than extended consciousness and has been nicely summarised by Jesse Prinz in an article attacking embodied theories of consciousness. Prinz writes that

> The claim that consciousness extends into the body is only marginally more plausible than the claim that consciousness leaks out into the

> world. We have never found any cells outside the brain that are candidates as correlates for experience. Such cells would have to co-vary with conscious states in content and time-course.
>
> (2008, p. 425)

In this passage, Prinz seems to take for granted a magical membrane of exactly the sort that we have seen Hurley attack earlier in this chapter. Nevertheless, the intuition he articulates is a powerful one. It has some purchase among philosophers otherwise sympathetic to the extended mind.

Clark (2009) took up the task of testing the strongest arguments he could find in the literature that defend extended consciousness. All the arguments he surveyed were taken from the writings of sensorimotor enactivists. We suggest that Clark is absolutely right that the Dynamical Entanglement and Unique Temporal Signature (in short, DEUTS) argument makes the best case for extended consciousness. We will argue that the DEUTS argument is best read as providing the bridge we are seeking from sensorimotor enactivism as an externalist theory of the qualities of phenomenal experience to extended consciousness as a how-explanation of phenomenal consciousness.

The DEUTS argument has two parts to it. The first part Clark labels "dynamic entanglement." We have already encountered the key idea in our presentation of sensorimotor enactivism. The key idea is that the qualities of experience are to be explained in terms of feedback loops that run through the body and out into the wider environment. Recall that it was mastery of SMCs that was said to account for sameness and difference in phenomenal experience. The exercise of sensorimotor skills, however, involves non-linear causal interactions among neural, bodily, and environmental elements. Sensory and motor channels enter into reciprocal causal influence based on the agent's interaction with the environment (Hurley 1998, 2010).

The second part of the argument Clark labels "unique temporal signature". The core idea is that in order to support experience, neural states must evolve over time in some specific way. This unique temporal signature cannot occur in the absence of the right environmental causes. To borrow an example from Noë, the flavour sensation one enjoys when drinking a glass of wine is in part the result of the agent-environment interaction that unfolds as the wine rolls across one's tongue (Noë 2004, p. 220).

Clark (2009) attempts to undermine the DEUTS argument by first attacking unique temporal signature. He argues that the external environment (such as the wine rolling over one's tongue) may well be required for driving neural activity. This fact on its own does not add up to an argument for anything external to the brain being a part of the "minimal machinery of experience" (2009, p. 983). At best it can be read as a claim about the qualities of experience as being essentially world involving. This much seems right.

But now consider unique temporal signature taken in conjunction with sensorimotor enactivism as an account of the qualities of experience. This combination gives us reason to think that for at least some phenomenal experiences, the internal neural processing that realises the experiences cannot be carved off or unplugged from the wider bodily and environmental setting in which they unfold. There is no carving off internal neural elements from bodily engagement with the environment. It is by exercising their mastery of SMCs that things have perceptual presence for perceivers. Taken together, dynamic entanglement + unique temporal signature imply that internal neural elements cannot be unplugged from the body and environment with which they are dynamically entangled. Some phenomenal experiences – those in which subjects are actively exploring the environment – are such that neural processes are on their own not sufficient for the occurrence of these experience. Only through feedback loops that unfold over time through the body and in interaction with the wider environment can subjects undergo these types of subjective experiences.

Clark attempts to further undermine DEUTS by putting forward the empirical hypothesis that the body acts as what he calls a "low-pass filter," thereby precluding the non-neural body and environmental elements from forming parts of the material realisers of conscious experience.

> A low-pass filter is a physical medium that allows low frequency signals through while reducing or blocking higher frequency signals . . . the extra-neural body . . . acts as a kind of low pass filter for signals coming from the environment. What this means in practice is that for phenomena that depend on e.g. the very fast temporal binding or processing of signals, the only locus in which such operations can (as a matter of fact) occur lies within the brain/CNS.
>
> (Clark 2009, p. 985)

Clark is here following up on a suggestion Chalmers made that "consciousness requires direct access to information on an extremely high-bandwidth" (Chalmers 2008, p. 6). Clark goes on to argue, based in part on the work of Wolf Singer and others, that the machinery of consciousness requires the fast temporal binding of sensory features through the synchronous activation of distinct neural populations. This synchrony correlates with attentional processes that allow information to pass from perceptual processes to working memory in a way that is required for the making of this information globally available to other consuming systems in ways that are required for consciousness. Insofar as the senses enable only a low-bandwidth connection to the world because of the body acting as a low-pass filter, this makes extended consciousness empirically unlikely. The timescale of information integration that consciousness requires makes neural processes the only likely vehicle for conscious experience.

Clark's low-bandwidth argument has the consequence that even if one were to accept sensorimotor enactivism as an account of the qualities of experience, extended consciousness would not be implied. The processing requirements for phenomenal experience make it the case that as a contingent matter of fact, consciousness must be fully realised inside the brain and cannot have realisers that extend out into the environment.

We have no doubt that Clark and Chalmers have a point when they insist that high-bandwidth connections matter for the realisation of consciousness. But is Clark right to claim that the body acts as a low-pass filter? Evan Thompson pointed out to one of us in personal communication at the time Clark's paper first appeared that "the time it takes visual stimulation to pass through the eye is a fraction of the time it takes neural system to build up correlated activity" (Thompson 2009, personal communication). Karina Vold makes a similar point:

> We know that our conscious states represent visual information that comes in from beyond our brain in a rapidly changing manner. Information about the surfaces of objects is transferred when light hits the eye, which is subsequently transferred to the brain. But the brain, which cannot transfer or receive information at the speed of light, slows this information processing down. So non-neural processes must be constantly reporting information back to the brain through the low pass filter Clark describes, at least as quickly as neural processes can operate.
> (2015, p. 21)

Chalmers has recently conceded this point to Vold (Chalmers forthcoming). However, he argues that there is nevertheless a related reason for remaining sceptical about extended consciousness. The neural underpinnings of consciousness, he argues, requires relatively direct access to information so that this information can be made available to processes of global control. Chalmers argues that while this condition is satisfied by internal neural processes, it is not satisfied in cases of extended cognition. This is because in cases of extended cognition, as Chalmers argues, "information must travel causal pathways from object to eye, from eye to visual cortex and from visual cortex to loci of control" (forthcoming, p. 10). But, having conceded that the senses provide high-bandwidth connections with the environment, why does Chalmers think such a connection could not deliver the information that is required for global control?

Chalmers spells out his reasoning as follows: Extended consciousness, he takes it, would require perceptual access to information located in the environment. Furthermore, the type of perceptual access he supposes is required would need to be conscious. So the information would first need to be consciously perceived, and then it would have to be made available for global control, requiring effectively two steps. But Chalmers is targeting an argument for extended consciousness that works along the same lines as the argument he made with Clark for the extended mind. This is not the argumentative strategy we have pursued.[5] We have argued for extended consciousness by combining sensorimotor enactivism with the DEUTS considerations that Clark attacks. Without the low-pass-filter argument, the argument against dynamical entanglement fails. Once DEUTS is back up and running, it seems to us that sensorimotor enactivism should be well placed to deliver extended consciousness. Can we avail ourselves of sensorimotor enactivism to argue for extended consciousness? In the next section, we will use Clark to raise a central problem for sensorimotor enactivists. This allows us to introduce another of the main characters onto the stage in section 2.5: predictive processing.

2.4 The weak spot in sensorimotor enactivism: dreaming and imagining

Sensorimotor enactivism claims that "experiences are essentially episodes of interaction between subjects and parts of the world" (Ward 2012, p. 734). A non-enactive view of phenomenal experience would, by contrast, treat the contribution of the environment to phenomenal experience as purely causal. When all goes well, the causal aetiology of an experience can be traced back to the environment. This, however, is a contingent (i.e., non-essential) fact about a given experience. As far as the subject of experience is concerned, things could appear or seem exactly as they do, while the causes of experience could be radically different. We have taken enactivism to claim that interaction with the environment makes a constitutive contribution to determining the phenomenal character of those experiences. Extended consciousness then fills in the details of how it is possible for interaction with the environment to make such a contribution to phenomenal experience.

Clark (2012) takes issue with the claim of the enactivists that experiences are essentially interactive and world involving. Clark tries to undermine sensorimotor enactivism by taking aim at its weak spot: its ability to account for dreaming and imagining. If experiences essentially involve relations to the world, how is the enactivist to account for similarities between perception, dreaming, and imagining? In the case of dreaming and imagining, the subject typically does not stand in a relation to whatever they are dreaming about or imagining. Yet these types of experience can share a good deal of their phenomenology with perceptual experience.

The sensorimotor enactivist must try to make sense of experiential episodes like dreaming and imagining while also trying to make good on the claim that experience *is* essentially interactive and world involving. To achieve this balancing act, the enactivist is forced to accept a view that treats dreaming and imagining as substantially different in kind from the kind of experiences one enjoys in daily waking life. But this sits uneasily with the experiential character of at least certain episodes of dreaming and imagining, which share a good deal of their phenomenology with perceptual experiences of the environment.

Clark's solution is to resist casting these experiential phenomena as "radically divergent states" (2012, p. 764). This means giving up on the enactivist claim that experiences are essentially world involving. The real task for any what-explanation of the qualities of phenomenal experience is to account for the similarities between perception, dreaming, and imagining while also making sense of how perceptual experience can be directly revealing of the world. Clark goes on to show how both of these desiderata are satisfied by subpersonal theories that locate the realisers of phenomenal experience exclusively inside the brain of the subject of the experience.

2.5 Clark and predictive processing

Clark appeals to the predictive processing account of the brain that he has done so much to develop for his subpersonal explanation of how the phenomenal experience may share a phenomenology with dreaming and imagining (Clark 2012, 2013, 2016). Predictive processing as Clark presents it has the benefit of making plausible a deep underlying unity between perception, dreaming, and imagination of exactly the type that we have suggested seems to make trouble for enactivism. It hypothesises that perceiving, dreaming, and imagining are simultaneous effects of "a single underlying neural strategy" (Clark 2014, p. 39). To consciously experience something in perception is for the brain to "deploy internal resources capable of endogenously generating those same activation patterns [as in dreaming and imagining]" (Clark 2014, p. 39). If in dreaming and imagining the brain deploys the same processes active during perception, then the underlying neural patterns supporting perceiving must somehow also support dreaming and imagining.

Predictive processing conceives of the computational processes realised in the brain's neural circuitry in terms of prediction error minimisation, reflecting the probability of sensorimotor input relative to an internal statistical, generative model. Generative models function to "capture the statistical structure of some set of observed inputs by inferring a causal matrix able to give rise to that very structure" (Clark 2016, p. 21). Thus, a generative model is a statistical model of how sensorimotor input are (most probably) generated, encoding prior probabilistic beliefs about which sensorimotor effects have which worldly and/or bodily causes. Each level of the brain's cortical hierarchy encodes a probability density function that aims to account for the prediction error signal arriving from the cortical level below (see, e.g., Rao and Ballard 1999; Hohwy et al. 2008; Hohwy 2013). (In Box 2.2, we provide further details on prediction error and generative models.)

This provides Clark with all that he needs (or so he claims) to reject the key claim of the sensorimotor enactivist: conscious experience is essentially interactive and world involving. He shows that predictive processing can explain such favoured enactivist examples as presence in absence. Our brains *do* have an implicit grasp of the kind of sensory effects that the agent would meet were it to engage in moving around the tomato (Seth 2014).[6] But, nothing about this demands that the specific realisers of such experiences are wide (Clark 2012). According to Clark, the generative model realised in the brain distils, on average and over time, information about causal-statistical regularities of the probable causes of its sensorium. As he says, when we are confronted with a scene involving a cat or a tomato, the brain's processing hierarchy "will relax into a stable state in which these higher-level patterns are recognised to be present" (2012, p. 762). Hence, Clark takes generative models to encode the kind of information about cats and tomatoes that we enjoy in conscious experience.

Predictive processing offers the prospect of a unified account of perceiving, imagining, and dreaming. All experiences – in perception, dreaming, imagination, and so on – are the result of the brain actively generating its own input in the form of top-down predictions and testing out those predictions against the world. We will have much more to say about the predictive processing account of consciousness over the course of this book. For now, we are showing how predictive processing challenges sensorimotor enactivism and, by extension, our argument for extended consciousness. Predictive processing offers a promising account of why the phenomenology of perception, dreaming, and imagining often strikes us as similar. It can also account for how experiences can be world revealing. Yet it also supports a view of the realising machinery of phenomenal experience as firmly placed in the brain's generative models. Thus, predictive processing would seem to present a serious obstacle standing in the way of the line of argument for extended consciousness that we have developed in this chapter.

Box 2.2 Generative models and prediction error

A prominent idea in computational neuroscience is that the brain performs approximate Bayesian inference (Friston 2010; Knill and Pouget 2004; Parr et al. 2018). The brain infers the *hidden* causes of its sensory observations. An intuitive example is the position (i.e., hidden variable) of a lamp causing a pattern of photoreceptor activation (i.e., sensory observations) in the retina. The idea is that the brain can use Bayesian inference to infer the probable position of the lamp given its sensory observations. As Parr et al. (2018) observe, to do this, the brain defines or instantiates two probability distributions (Figure 2.2).

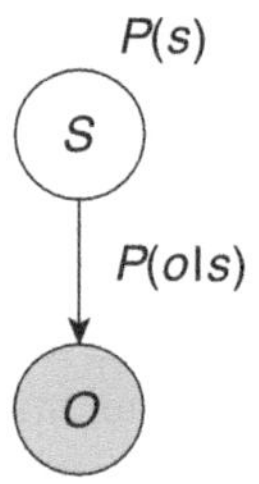

Figure 2.2 Simple illustration of a generative model comprising a set of joint probability distributions

Source: from Parr et al. 2018, p. 3; used with permission from the Wellcome Trust

This figure depicts a simple generative model comprising two probability distributions: the prior probability distribution of the causes, $P(s)$, and a likelihood distribution that specifies how hidden variables or causes determine sensory observations, $P(o|s)$. Jointly, these probability distributions describe the processes through which sensory observations are generated under approximate Bayesian inference (Parr et al. 2018).

The connection between generative models and prediction error comes via the predictive coding scheme in theoretical neurobiology. Predictive coding expresses the idea that the brain makes use of the generative model to infer its own perceptual hypotheses – its "best guesses" about the causes of its sensory observations (Rao and Ballard 1999). The difference between any specific prediction (e.g., that the lamp is in position L instead of N) and the actual sensory data is compared. Prediction error is therefore the distance between the prediction and the actual sensory observation. Prediction error can be used to update or refine predictions about the causes of sensory observations.

2.6 Summary

In this chapter, we used sensorimotor enactivism to provide an externalist theory of the qualities of experience. We introduced the DEUTS argument to support an inference from sensorimotor enactivism to extended consciousness. Clark's argument seems to show that the phenomenological similarities between perception, dreaming, and imagining undermine sensorimotor enactivism. If there are cases of conscious experience that are phenomenologically just like perception but that unfold in the absence of actual bodily activity in the world, then the claim that some experiences are essentially world involving loses its plausibility.

In the next chapter, we will develop a reply to Clark by arguing that imagination and dreaming are derivative cases that inherit their phenomenology from the dynamics inherent in agent-environment couplings underlying extended consciousness. In imagination, for instance, perceptual experiences are reused through episodic memory to imagine an object in its absence (Hutto and Myin 2017, ch. 8). If this response works, it would leave open the possibility to continue to account for the quality of phenomenal experience along the lines that the sensorimotor enactivist proposes. The sensorimotor enactivist need not deny that imagination and dreaming share a common phenomenology with perceptual experience (though there are also significant phenomenological differences).

Notes

1 See also Thompson and Varela (2001) and Cosmelli and Thompson (2010).
2 The "plug and replug" terminology we borrow from Hurley (2010) and her discussion of supervenience thought experiments. See also Hurley (1998, ch. 7).
3 Noë argues that adaptation to the device occurs in adults that do not exhibit the same plasticity as the newborn ferrets we just discussed. Moreover, the active user of the device can begin to have quasi-visual experiences fairly rapidly after just a few hours of use, too soon for this change in experience to be due to structural changes in the brain that reroute the tactile information to different parts of the brain. These arguments are somewhat complicated by evidence of cross-modal plasticity in congenitally blind users of sensory substitution. For a discussion of these findings, see Kiverstein et al. (2015). To avoid these kinds of complications, we can run the argument for sighted users of sensory substitution who are blindfolded. For further discussion and pushback against sensory substitution–based arguments for extended consciousness, see Wheeler (2015b).
4 We are not sure whether this is how Kevin O'Regan thinks about SMCs. To us, his work seems ambiguous on this point, but it is at least in the spirit of O'Regan and Noë's talk of the perceptual systems as being attuned to SMCs in their 2001 article.
5 In Chapter 4, we will return to Chalmers's objection to extended consciousness. Chalmers's objection is premised on the common-sense idea that perception and

action function as interfaces where minds meet up with the world. Perception and action thus can serve as the boundary of the mind. In Chapter 4, we will show how this common-sense idea can be given a formal treatment that raises problems not only for the hypothesis of extended consciousness but for the extended mind more generally.

6 We discuss Seth's account of presence in absence in Chapter 5.

3 Extended dynamic singularities – models, processes, and recycling

3.1 Introduction

The previous chapter surveyed a number of arguments for extended consciousness. We paid special attention to the argument from sensorimotor enactivism. We suggested that the DEUTS argument may provide sufficient reasons for thinking that continuous sensorimotor engagement with the niche – engagement composed of dynamic loops breaking across brain, body, and world – are sometimes necessary for conscious experience. Such a line of argument has, however, not gone unchallenged. We saw how Clark used predictive processing to apply pressure to any such conclusion. If there are cases of conscious experience that are relevantly similar in their quality and richness to online perception but that unfold in the total absence of bodily activity in the world, then the DEUTS argument for extended consciousness – or any set of claims along those lines – is not plausible.

Predictive processing seems to be an obstacle that stands in the way of any argument from sensorimotor enactivism to the thesis of extended consciousness. Under predictive processing, organisms realise (or, better, their dynamics instantiate) a generative model. A generative model is a statistical mapping from hidden (external) causes to input under a statistical model – that is, under a set of joint probability distributions consisting of prior probabilities (the recognition distribution) and the posterior probabilities (the variational distribution) (see Chapter 2, Box 2.2). Less technically, one can think of generative models as encoding information about statistical regularities, reflecting certain expected regularities in the processes from which an organism's sensory data are generated. Crucially, once such information has been gained, it appears that bodily activity in the world is no longer essentially involved in the realisation of conscious experience, because the information needed for expressing conscious experience is encoded internally in the generative model.

Many researchers in the predictive processing camp take the function of generative models to be the generation of virtual or mock sensory input. The internal virtual models that explain offline episodes of conscious experience – in dreaming and imagining, say – also explain the character of online conscious experience in embodied activity in the world. We call this the inferred fantasies view (Kirchhoff 2017). The inferred fantasies view takes the function of the generative model to be the performance of a form of controlled guesswork that enables the brain to predict sensory information. In this view, what we perceive is not driven by direct sensorimotor engagement with the world but determined entirely by hierarchical generative models, constructing and generating patterns of sensory stimuli for themselves "in fantasy." As Hobson and Friston put it, generative models "[furnish] a theatre in which fantasies can be rehearsed and tested against sensory evidence" (2014, p. 6).

There is however, a complication with the inferred fantasies view. It implies that there must be some endpoint at which the brain can sit back and rely on its own simulated predictions to determine conscious experience. But there is no such endpoint during embodied activity in a dynamically changing environment. This embodied activity is embedded in a network with internal and external orbits comprising a single dynamical system (tenet one of the third wave: see Table 1.1). Any dynamical singularity will have internal and external orbits that are fluid and changing. So in active, bodily engagement – that is, in active inference – there will be no stage at which an agent can rely exhaustively on their neurally realised generative model. The hierarchical generative model is necessary but not sufficient for generating conscious experiences of the world. What goes wrong with the inferred fantasies view can be brought out, or so we will argue, by reconsidering the DEUTS argument for extended consciousness but this time through the lens of predictive processing. However, before we get to this argument, we must first provide a more detailed statement of the inferred fantasies view.

3.2 Inferred fantasies

The inferred fantasies view claims that our conscious experiences are the consequence of underlying (subpersonal) generative models and that generative models are formally equivalent to virtual reality models. Hence, an experience is an "inferred fantasy about what lies behind the veil of input" (Paton et al. 2013, p. 222) or a process of "controlled hallucination" (Clark 2013, p. 25) constructed to "keep the sensory input at bay" (Hohwy 2013, p. 137). According to its main proponents, this means that the brain is best conceived as a kind of virtual reality machine, where "predictions (fantasies) are generated in a virtual model of the world and then tested against sensory reality" (Hobson and Friston 2014, p. 9).

We can bring out this view informally via the following example: Let us say you walk into a hardware store and ask, can I get *four candles*? Now, were the shopkeeper to hand you *fork handles*, you might well be surprised. What explains the fact that the shopkeeper seemingly heard you saying "fork handles" when you asked for "four candles"? Informally, the explanation is that our brains assemble models of the world and keep modifying or updating the parameters of these models. Our brains are continuously busy, fine-tuning the posterior probabilities of their predictions (i.e., the probability of these predictions in light of new sensory data). This means that what the shopkeeper perceives – what she experiences – is her brain's model of the most probable prediction given her sensory observations. In this sense, one could say that her perceptual experience is a fantasy that may or may not correspond with an external state of affairs (i.e., with the true causes of her observations). The shopkeeper does not simply hallucinate. Nor does she come to experience the world willy-nilly. The idea is that she meets her sensorium with posterior beliefs, some of which get to determine her conscious experience.

We can describe this scenario in more formal terms as one of approximating Bayesian inference. The function of the shopkeeper's generative model is to infer the probability of a prediction or hypothesis, *H* (the event that the customer said – e.g., "fork handles"), given some evidence, *E* (the acoustic sound). Her generative model must do work to infer the posterior probability (or posterior density). The formal notation is written $P(\mathrm{H}|\mathrm{E})$. On the assumption that the shopkeeper's brain operates in accordance with approximate Bayesian inference, it follows that the internal configuration of neuronal states changes over time in ways that approximate the posterior density distribution, thus accounting in a probabilistic way for the causes of the evidence. Approximating the posterior density is possible because the posterior density is equal to the conditional probability of the evidence given a hypothesis, $P(\mathrm{E}|\mathrm{H})$, multiplied by the prior probability of a hypothesis, P(H), divided by the marginal likelihood of the evidence, P(E). Thus, via approximate Bayesian inference, the shopkeeper's brain allows her to update the posterior under a generative model by on average following Bayes's rule.

There is something correct about this. This is the insight that what we come to perceive depends essentially on processes of probabilistic inference (see Box 3.1 for why some might worry that our agreement with this conflicts with our agreement with sensorimotor enactivism).

Box 3.1 Inference and sensorimotor enactivism

One might worry that a Bayesian inferential theory of perception is likely to cause trouble for our endorsement of sensorimotor enactivism in the previous chapter. This is an important worry that we will have cause to return to at different places in our book (particularly in Chapter 5).

Sensorimotor enactivism claims that the qualities of (some) perceptual experiences depend on an agent's implicit mastery of SMCs – the counterfactual relations between movement and sensing. Implicit mastery of SMCs can be cast in predictive processing terms as predictions about what sensations the agent would encounter were it to move in such and such ways (Seth 2014, 2015a). However, as we will discuss in Chapter 5, Seth's account of mastery of SMCs appeals to the information encoded in a generative model. This has internalist consequences that spell trouble for any argument for extended consciousness. We respond to this challenge in detail in Section 5.1.

A second worry might be that if one assumes that perceptual experience is tied up with probabilistic inference, then one cannot also hold onto the dynamic singularity view endorsed by sensorimotor enactivists. To repeat, sensorimotor enactivism holds that the nervous system sustains the organisation of the organism as a mobile unity, implying that agents meet their niche conditioned on their own sensorimotor terms (Thompson 2007, p. 244). This means that embodied sensorimotor activity is what enables perceptual experience. Experience is realised in recurrent and reciprocal interactions between motor behaviour, sensory input, and the implicit mastery of counterfactuals between moving and sensing (Thompson 2007, p. 256). The worry is that this view runs counter to the claim that experience is the outcome of probabilistic inference. We show in Section 3.2.1 of this chapter that this objection has no real force, by distinguishing between internally realised (relative to organismic boundaries) generative models, on the one hand, and environment-involving generative processes, on the other.

The first assumption of the inferred fantasies view that we wish to emphasise is the claim that there is no substantial difference in the character of offline and online species of experience. There are many defenders of this view. Revonsuo (2015), for example, argues that current evidence from empirical dream research supports the idea that dream experience is qualitatively similar to or sometimes even indistinguishable from waking experience. In other words, dreaming experience is as rich, vivid, and varied as waking experience is. This claim runs counter to those that argue that there is a difference between experience in dreaming and experience in waking life. For example, Noë (2004) argues that this (alleged) difference is due to waking experience being realised in embodied activity in the world. But Noë's claim is unlikely to persuade the defender of the inferred fantasies view. They will try to relax the initial equivalence claim concerning online and offline conscious experience and still discount Noë's (2004) conclusion. Clark (2014) makes precisely this move by arguing that the processing of sensory information involves generative models inferring the causes of sensory observations. The information served up by the world matters, as it provides a stream of data that the brain's self-generated predictions have to try to match. Dreaming and imagining are different in this respect given that these are examples of conscious experience that do not depend on an evolving sensory flow of information elicited from bodily or worldly causes. There is more to say about this move by Clark, as we will see below. But the main point for now is that the difference in driving signal – but not in the realising generative model – explains any experiential difference that there may be between vivid perception and non-vivid dreams and imaginings. This view is not specific to Clark. We find the same point in Hohwy (2013):

> This fits with the idea that conscious experience is like a fantasy or virtual reality constructed to keep sensory input at bay. It is different from the conscious experience that is truly a fantasy or virtual reality, which we enjoy in mental imagery or dreaming, because such experiences are not intended to keep sensory input at bay.
>
> (p. 137)

The second assumption of the inferred fantasies view is that conscious experience is not essentially tied up with bodily and worldly activity, even if some of our everyday experience would not be possible without at least initial influence from the body and world. The philosophical implication of the inferred fantasies view is, however, that once encoded, bodily and worldly regularities play second fiddle to internally realised prior beliefs in the generative model. It is this point that drives proponents of this view to say that our conscious experiences in dreaming are elicited by the same processes as in waking perceptual and bodily experience.

This unifying view of the processing underlying dreaming and perceiving – offline and online modes of experience – is sometimes referred to as the simulation theory of conscious experience.[1] There are different versions of the simulation view in the literature. Hobson and Friston (2014) recently developed one species of a simulation view premised on predictive processing. They argue that the simulated virtual reality model involved in dreaming experience is equivalent to the generative model induced in waking experience. This is premised on the claim that what we experience is not, strictly speaking, determined by sensory observations but rather by the brain generating its own sensory stimuli (see also Bar 2007). This means that conscious experience can arise in the absence of bodily and worldly activity. Hobson and Friston take this to imply that conscious experiences "are literally fantastic" (2014, p. 10). There is therefore no reason to think that the kind of conscious experiences we enjoy while awake are realised in anything other than the kind of generative models activated during sleep. Conscious experiences are inferred fantasies realised by neurally realised generative models both in dreaming consciousness and in waking consciousness.

Clark agrees, as we saw in the previous chapter. He says that "Perceivers, like us, if this is correct, are inevitably potential dreamers and imaginers too. Moreover, they are beings who, in dreaming and imagining, are deploying many of the very same strategies and resources used in ordinary perception" (2012, p. 764).

We agree with defenders of the inferred fantasies view that consciousness can with good grounds be described as a process of approximating Bayesian probabilistic inference. We explain further why we think this hypothesis is on the right track for explaining conscious experience as the book proceeds. Yet something can nevertheless be shown to have gone wrong with both assumptions of the inferred fantasies view we have just outlined, as we now proceed to show.

3.3 The DEUTS argument reconsidered

Why should we question the inferred fantasies view? Our reasons for doing so stem from the DEUTS argument for extended consciousness.

The first is that the inferred fantasies view renders the external parts of the dynamic singularity – the ongoing web of extended dynamics with feedback loops running across internal and external orbits – inessential to the realisation of conscious experience. But the claim that the ongoing extended dynamics are not necessary for online perceptual experience comes at a price. It overlooks the important observations that the "flow of sensation (bound in . . . a constant circular causal embrace with the flow of action) is predictable just to the extent that there is a spatial and temporal pattern in that flow" (Clark 2016, p. 171). Later in the same discussion, Clark notes that this "pattern is a function of properties and features of the world and of the needs, form, and activities of the agent" (Clark 2016, p. 171). Hence, the flow of sensation during online perceptual experience is inescapably bound up with and embedded in an ongoing extended dynamic process.

We argue that the generative model can be conceived as realised not only in the brain but also in an extended dynamic singularity: in a continuous dynamic network with internal and external feedback loops (Hurley 1998). Our argument draws its inspiration from third-wave extended mind. It also finds a natural home in radical enactivism, which states that the vast majority of what we do and what we experience is realised by continuously unfolding embodied interactions and engagements with the world (Hutto et al. 2014; Kirchhoff and Hutto 2016; Kiverstein 2016).

Our second reason for questioning the inferred fantasies view is that it puts the cart before the horse in view of the relation between offline and online conscious experience that it implies. Proponents of the inferred fantasies view argue that if the generative model underlying purely offline experience such as we have when dreaming can also fully account for experiences had during online perception, then the external orbits play no essential role in realising conscious experience. By contrast, we contend that simulating online experiences offline – through neurally realised generative models – should be understood in terms of reuse or functional redeployment (Anderson 2014). The brain *reuses* some of the same posterior probabilities generated and attuned by an agent to recurrent agent – environment dynamics in online perceptual experience. Simulating experiences offline thus piggybacks on online perceptual experience realised in an extended dynamic singularity. This argument owes much to Hurley (2010).

We deal with each of these arguments in turn.

3.3.1 Dynamic singularities, embodied generative models, and generative processes

We have seen above how some proponents of the inferred fantasies view make the following argument. They argue that if the quality of our experiences in dreams is indistinguishable from that of our experiences while awake, this lends support for the claim that the brain, apart from its coupling to the world, is sufficient on its own for conscious experience. They reason in the same way for imagining. If the character of our experiences when imagining is the same as the character of our experiences in online embodied engagement with the world, then the necessary and sufficient realisers for conscious experience are those involved in imagining: message passing in a neurally realised generative model. But (as we saw above), this claim overstates the case for internalism. It is unlikely to be correct that the experiential character of experience is in each and every case the same across offline and online modes of experience. Clark (2014) agrees. He argues that there is a distinct difference in what he calls the experiential signature of perception and imagination. This is, of course, not a claim unique to Clark. Hume acknowledged these important phenomenological differences when he argued that perception has a liveliness, force, and vivacity that is subdued if not entirely missing in imagination (Hume 1738–1740).

Clark argues that part of the reason for this difference in phenomenology stems from the fact that online perceptual experience is partly driven by a pattern of the sensory signals, whereas this need not be the case for offline imagination. It is this difference – and resulting differences in the processing of the signals (or their absence) – that "is suggestive. . . of a possible explanation for the experiential asymmetry between perception and (non-vivid) imagination" (2014, p. 18). If Clark is on the right track, Revonsuo's (2015) experiential equivalence claim will not work. At least, it will not straightforwardly imply his version of the simulation view, which claims that the processes taking place within the brains of individuals when they are asleep always suffice for the rich world-revealing conscious experiences that people ordinarily enjoy.

Acknowledging that this difference in the phenomenological character of online and offline experience is due to patterns of sensory signals being shaped and orchestrated by the environment and the "needs, form, and activities of the agent" (Clark 2016, p. 171) opens the door once again to a DEUTS-style argument for the thesis of extended consciousness. That is, one is quite naturally led to consider the role of embodied action or active inference in the overall story about predictive processing.

The usual way of presenting active inference is as a distinct strategy for prediction error reduction to perceptual inference (Hohwy 2012). Perceptual inference is cast as a process of minimising prediction error by updating the posterior distributions of the generative model. Active inference is associated with the process of minimising prediction error by keeping the posterior distributions of the generative model stable and instead seeking to revise and alter the sensory stream to fulfil the predictions of the generative model.

Everyone agrees that without active inference, an organism would not be able to minimise prediction error (Hohwy 2012). This is also why Friston (2018) says that active inference (aka the free energy principle) "is not just about making the best (Bayesian) sense of sensory impressions of what's 'out there.' It tries to understand how we sample the world and author our own sensations" (2018, p. 22). Active inference has the consequence (or so we will argue) that what we experience in perception is partly realised by what we do – by how we act – just as sensorimotor enactivism contends.

This is a familiar theme in the predictive processing formulation, which casts perception and action as being tightly coupled to one another (Clark

2013, 2016; Friston 2011; Hohwy 2013). But perception and action are not merely closely coupled. Action underwrites perception in the sense of being a precondition for perception. In active inference, perception turns out to be an embodied activity because perception and action are constitutively interdependent (Hurley 1998). Active inference, we argue, implies an enactive view of online perceptual experience according to which the latter unfolds in an extended dynamic singularity. Clark (2016) comes close to reaching a conclusion along these lines when he writes that "Rather than aiming to reveal some kind of action-neutral image of an objective realm, prediction-driven learning delivers a grip upon *affordances*: the possibilities for action and intervention that the environment makes available to a given agent" (2016, p. 171).

None of this is to deny that internally generated predictions are necessary for conscious experience. Indeed, were an organism to make no predictions of not-yet-actualised consequences of its own sensorimotor actions, and were it to receive no subsequent error or feedback signals, but only flows of environmentally driven signals, it would most likely lack any kind of conscious experience and would certainly be unable to engage with affordances (Hurley 1998; Gallagher 2018). For an organism to be conscious, its generative model must exhibit what Seth (2014, 2015a) and Friston (2017) refer to as *temporal thickness* or *counterfactual depth* (c.f. Kiverstein 2018). This means that any agent that is able to grasp the consequences of its future actions must have a generative model that can entertain probable consequences of its not-yet-actualised sensorimotor activities over different timescales.

Associating conscious experience with inferential processes "imbues consciousness with a hierarchical aspect" (Hobson and Friston 2014, p. 11). Active inference is premised on generative models with temporal thickness enabling individual agents to act in such a way as to maintain themselves within certain expected states in the long run (Hohwy 2016).

Our argument for associating active inference with sensorimotor enactivism requires thinking further about the conditions necessary for agents to maintain themselves within long-run expected states. A simple condition is that organisms must act in a way that maximises the probability that they will remain alive over time. One should not confuse the claim that active inference imbues organisms with temporal thickness with the claim that action plays second fiddle to internal processes predicting the counterfactual outcomes of acting in the world. It is action that gives an organism a grip on its environmental affordances, enabling an organism to remain within a set of limited species-typical states and thus to avoid crossing terminal phase boundaries (Bruineberg et al. 2016; Kirchhoff et al. 2018). An example of a terminal phase boundary is a cliff edge. On one side of the boundary, a flightless organism will retain its organisational form and integrity. On the other side, it is highly unlikely that it will do so. The smart thing to do for the agent is to sample among different options for how to act and select the option that has the best (predicted) evidence or the least (expected) prediction error. The options sampled from are intuitively probabilistic and future oriented. Hence, active inference enables an organism to "free" itself from its proximal sensory signal by engaging in inferences about probabilistic future states and acting to minimise the uncertainty associated with being in those possible future states. Systems with deeper temporal structures will fare better when it comes to predicting the counterfactual outcomes of their actions.

Temporal thickness matters when explaining the difference between the qualities of online perceptual experience and offline, non-vivid imagination. Clark (2014) gives the following example:

> [S]uppose you are looking for an object on a crowded surface. You expect to see it somewhere, but you are not sure. Your brain must temporarily increase the weighting [i.e., the reliability] in the fine spatial information carried by the sensory signal. That way, you don't simply mistakenly see it there (at such and such a location) just because you are expecting to see it somewhere. To match the driving signal with a top-down prediction here demands accounting for the sensory signal in great detail, all the way down (as it were).
>
> (p. 18)

Contrast this with non-vivid mental imagery, which may be relying on processing at higher levels of the generative model. The example that Clark has in mind is as follows:

> Thus, compare the case where you are asked to imagine your walk to work. Here, early (closer to retinotopic) stages of the processing hierarchy can be allowed substantial leeway, and need not be forced to settle into one interpretation or another. It seems plausible. . . that under such conditions one might experience the self-generated imagery as fuzzier and less distinct than online perception.
>
> (2014, p. 18)

In both cases, predictions are involved. In the latter case, Clark suggests that the processes involved in offline mental imagery need not pay as much attention to the driving signals from the environment. Indeed, one interpretation is to say that during offline simulations of an everyday perceptual experience – of walking to work, say – the generative simulations are making indirect use of the kind of extended dynamics that embeds an agent's activities during waking perceptual experience. The former case privileges how the flow of input is sculpted by the embedding situation. This suggests, prima facie, at least, that the external parts of the extended dynamic are part of what explains the difference in the experiential character of online perceptual experience and offline, non-vivid imagination.

But we can build on and deepen this point. The idea that external factors of extended dynamic processes in part explains the difference in phenomenal character between offline and online conscious experience speaks to two key issues. First, it relates to the distinction sometimes made between "having a model" and "being a model" (Friston 2011; Friston et al. 2012a). Second, it points to an important but often overlooked distinction between a "generative model" and a "generative process."

First, predictive processing considers organisms as close to optimal models of the causal-statistical patterns or regularities that make up an organism's species-typical environment. This optimality is achieved by minimising prediction error (on average and over time), which bounds the evidence for each individual – that is, its model. One way to read this claim is to take it to imply something akin to a Pac-Man-style scenario. In the arcade game, the player has to manoeuvre Pac-Man through a maze, the aim of which is to get Pac-Man to eat all the Pac-Dots (or biscuits) populating the maze. Success then leads to the next level, the next after that, and so on. By rough analogy, evolution has set up organisms such that they, on average and over time, distil (i.e., extract) statistical regularities of their niche and therefore come to embody such regularities in their generative model. Much like Pac-Man, the biological system moves around extracting and consuming structures from its environment. This is how it learns about its niche. Proponents of the inferred fantasies view think of the body-plant and wider context as important for acquiring and updating the parameters of a neurally realised generative model. However, once the organism has distilled the statistical regularities of its environment, it can throw that environment away and rely instead on its internally encoded generative model of the world. It is a view like this that we see in Hohwy: "The mind can then be understood in internalist, solipsistic terms, throwing away the body, the world and other people." Hohwy endorses a view that he finds articulated in Shapiro: once brains are given their input, "the rest of the world makes no difference to them" (Shapiro 2011, p. 26, 2016, p. 265).

The Pac-Man model is not the right way to think about the relation between the organism and the world.

In active inference, individuals must constantly work to maintain a grip on the random fluctuations in the dynamics of their environments. The fluctuations do not reside or disappear but are constantly forming and reforming, even if in slightly different ways. This means that the minimisation of prediction error is ongoing during online perceptual experience given that the embedding environment is dynamic and always shifting. Hence, agents must constantly adjust their internal dynamics to the continuously changing dynamics of the environment, in particular to its possibilities for action. One does not distil affordances in the way that Pac-Man distils Pac-Dots. Indeed, there is no *endpoint* at which an organism can just rely on its internal generative model independent of the niche in which it is situated. Defenders of the inferred fantasies view might regard the continuous maintenance of generative models by dynamics of the environment as nothing but the necessary causal backdrop against which the brain's prediction error minimisation processes operate in the here and now. But, as Hurley (2010) rightly pointed out, one should not "assume that extended tuning and maintenance processes cannot be part of the sought-for explanation of how experience works" (2010, p. 142).

This suggests (minimally) that one needs to re-evaluate how to think about the generative model. The organism does not have a generative model as though there were two separate things – the organism and a generative model housed inside of its head. Instead, we ought to think of agents as *being a model* of their environments, where being a model consists in being an extended dynamic singularity. Consider, in this light, how Friston et al. (2012b) define a generative model:

> We must here understand "model" in the most inclusive sense, as combining interpretive dispositions, morphology, and neural architecture, and as implying a highly tuned "fit" between the active, embodied organism and the embedded environment.
>
> (2012b, p. 6)

Friston has written somewhat enigmatically that an "agent does not have a model of its world – it is a model" (2013, p. 213; see also Bruineberg et al. 2016; Kirchhoff 2016, 2017). We unpack this embodied conception of the generative model as follows. Every biological agent can be described as a probability distribution – a hierarchically organised probabilistic model conditioned on the sensory, physiological, and morphological states that are highly probable given the life it leads and the eco-niche it inhabits (Friston 2010, 2013). Life can be described as a process of inferring a probabilistic model that identifies the bodily states that the organism has a high probability of finding itself in. To maintain its own organisation in its sensory exchange with the environment, an organism will need to keep the discrepancy between the predictions of its model and what actually ensues to a minimum. It will need to act so as to avoid "surprise," understood technically as the negative log probability associated with a bodily state or informally as the improbability of the organism finding itself in such a bodily state on average and over time (for a detailed discussion, see Kirchhoff and Robertson 2018; Kiverstein 2018).

The generative model should therefore be interpreted as instantiated by the agent as a whole. In other words, it is not something that one can abstract away from the phenotypic traits of an organism, because it is those traits, including states of its local niche, that instantiate such a model. In aiming to continuously improve the fit of its model to the sensory observations it gathers in its interactions with the environment, the agent is aiming to maximise the evidence for its own existence. That is, the agent generates changes in its own active and sensory states in ways that maximise the evidence for its own continued existence. Under active inference, the only self-consistent prior is that the actions undertaken by organisms will minimise prediction error (Mirza et al. 2016). Active inference provides a "formal, information theoretic framework within which to explain the constitutive coupling of the brain to the body and the environment" (Allen and Friston 2016, p. 16). For these reasons, we think of the generative model as the entire embodied organism.

Second, we suggest it is important to distinguish between a generative model and a generative process. As we have seen, active inference implies that a generative model, comprising two joint probability distributions, is realised by an embodied sampling of the world – that is, through embodied activity. This means that we should understand the generative model as a statistical model in which information is expressed and orchestrated through a generative process. Generative processes couple agent and environment through active and sensory states. Crucially, it is the generative process that makes possible the generation of sensory observations, underpinning the claim we made above that perceptual experience is realised in action. Specifically, the generative process is the pragmatic and epistemic actions of an embodied agent that couples environmental dynamics to the dynamics of the agent, where the coupling is what enables an agent to reduce uncertainty. The generative process thus allows an agent to "disclose information through exploration that will enable pragmatic action – actions performed for pragmatic ends – in the long run" (Friston et al. 2015a, p. 2).

It is through the generative process that the generative model is realised. The generative model must be distinguished from the generative process because the actual causes of sensory input depend on pragmatic and epistemic actions – that is, the generative process. Pragmatic and epistemic actions in turn depend on inference – that is, on the generative model. The generative model selects (i.e., infers) ways of acting (policies) that have the greatest evidence or the least expected prediction error. The generative process enacts these specific ways of acting. Active inference should therefore be understood as directed towards learning priors – that is, probability distributions – that inform action and the effects of action on the hidden causes generating sensory observations. What is inferred are the outcomes of specific sensorimotor engagements with the environment. The inferential process comes to an end with a posterior distribution that specifies that a particular action be performed. However, it is the action that elicits the sensory observations that the organism's inferential processes must get a grip on. The main work of the generative model is therefore to control the coupling of the organism to its embedding environment via action, which in turn completes the perception-action cycle (Friston et al. 2017). We can schematise this in Figure 3.1.

The prior probability distributions that the generative model maps can be thought of as the agent's sensorimotor abilities. The generative process situates the dynamics of the generative model in the environment. It also has the function of securing the agent's grip on the affordances offered up by the environment conditional on its sensorimotor and other skillful capacities. We thus agree with Gallagher and Allen that active inference "is not something happening [exclusively] in the brain and is not just providing new sensory input for the brain; it's what the whole organism does in its interactions with the environment" (2016, p. 12). In active inference, the generative model and

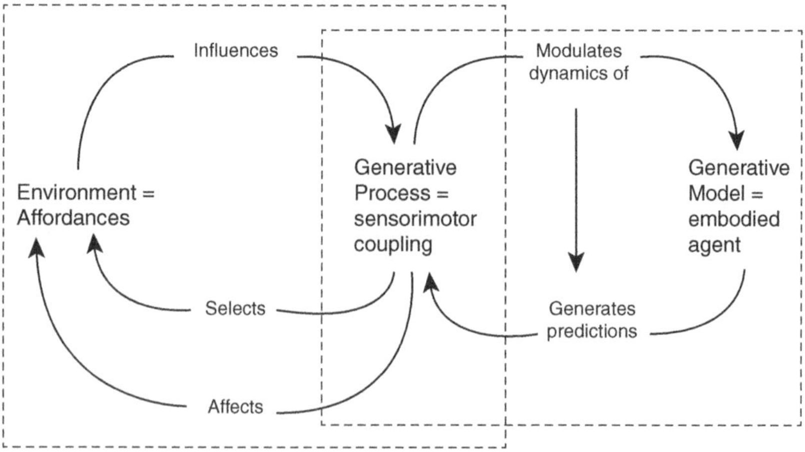

Figure 3.1 Illustration of the generative model (i.e., embodied agent) coupled to the environment via the generative process, and subsequent influences between the quantities involved in the minimisation of prediction error

Source: modified from Chemero 2009, figure 7.1

the generative process are locked together in a continuous reciprocal loop (Ramstead, Kirchhoff and Friston, under review). This means that generative models coupled in active inference to generative processes are uniquely equipped to make use of the properties of an organism's embodiment and associated species-typical environment.

3.3.2 Reusing extended dynamics in offline experience

We start this subsection with a potential objection. The inferred fantasies view was introduced by providing an example of a shopkeeper who fails to get a grip on the actual causes of their sensory observations, thus incorrectly enacting a prediction of "fork handles" even though what was asked for was "four candles." Does this not suggest that what she experiences is something like a fantasy after all – something that is produced solely by internal processes in her brain? Responding to this objection will help us to bring out a deeper reason for why it is a mistake to construe conscious experience as realised by nothing but neural processes.

We have been arguing that active inference does not take place in isolation from the environment but instead unfolds in the process of the agent's coupling to their environment. The generative model functions to regulate and control the agent's coupling to the environment. One cannot simply pull action out of the embedding environment, because actions unfold in and are shaped by the patterned regularities that comprise an individual's niche (Roesptorff et al. 2010). It is correct to say that the shopkeeper hears what she does given the specific posterior distributions of her generative model. But what explains the continuous maintenance and updating of particular posterior distributions? The obvious yet overlooked answer is that posterior distributions are maintained on the basis of the generative process, which in turn is influenced by the patterned regularities of cultural practices (Figure 3.1).

The proponent of the inferred fantasies view might insist that the constitutive basis of the shopkeeper's experience is her enbrained generative model. What would be the implication of denying this internalist neurocentric view of experience? Perhaps it might be the equally extreme claim that it is the cultural practices that shape and constrain the development of the shopkeeper's posterior distributions. It is tempting to see our view as one that would prefer the latter – giving full credit to that external part of what we are referring to as an extended dynamic singularity. But doing so would misinterpret our argument. Instead, one should explain the realisers of the shopkeeper's experience as comprising the constraints that arise out of patterned cultural practices; the sensorimotor constraints that govern embodied interaction with the environment and that are reflected in the generative processes; and their embodied generative model. This is the wide realisation base of the shopkeeper's experience (Kirchhoff 2015c).

This brings us to the point we wish to make in this subsection. The inferred fantasies view insists that the internal simulations produced during offline conscious experience are sufficient to explain experiences enjoyed in online perceptual activity. Assuming that this is correct, the inferred fantasies view gives the inner simulated models "explanatory independence from extended online dynamics" (Hurley 2010, p. 142).

The inferred fantasies view goes for an extreme, full blanket claim: all experiences are the result of an internally instantiated generative model. But why not allow for the more ecumenical position that processes that enable conscious experience can be offline in some cases and part of an extended dynamic singularity in other cases? This is precisely the kind of position that one ends up with by stressing that online perceptual experience is partly grounded in the generative process, whereas this is not the case in offline perceptual experience, which is not continuously modulated by generative processes coupling the generative model to the agent's environment.

The inferred fantasies view casts the virtual realities of internally instantiated generative models as primary or prior to the kind of generative models enacted during online perceptual experience. The inferred fantasies view holds that models encoded in upper layers in the cortical hierarchy are able to predict the dynamics at the level immediately below, and so on all the way down the brain's hierarchy. As Clark (2016) puts it, "Since this story applies all the way down. . . that means that such systems are fully capable of generating 'virtual' versions of the sensory data themselves" (p. 93). However, if online perceptual experience is constituted by a dynamic singularity comprising the generative model, the generative process, and elements of the environment, this gives us grounds for treating online experience as having primacy, not dreaming and imagining. Imagining and dreaming are the products of internal generative models recycling posterior distributions generated, attuned, and maintained by unfolding generative processes during online perceptual experience.

Of course, this is to concede the point that some varieties of phenomenal experience – those that occur in dreams and in some forms of imagination – are constituted by the generative models decoupled from the environment here and now. Conceding this is a virtue. It allows us to explain what imagining, dreaming, and perceptual experience have in common, phenomenologically. All varieties of phenomenal consciousness are generated by processes that function to minimise long-term prediction error. We can also account for the important differences in phenomenology that accompany these different types of phenomenal experiences (as discussed in Section 3.1). Finally, our account allows us to consistently claim that perceptual experiences of the world are realised by extended dynamic singularities, without our needing to maintain that the same is true across the board for all types of phenomenal experience.

3.4 Summary

We have argued that some kinds of conscious experience are grounded in an extended dynamic singularity. We made this argument by distinguishing between the generative model and the generative process, the latter coupling the agent to the environment and vice versa. This provides an account of prediction error minimisation as more than perceptual inference. It grounds perceptual inference in action. We did not claim that this holds for all and every kind of conscious experience. As we have presented predictive processing here, it implies that there is a non-trivial difference in the constituting base of perceptual experiences as compared with the modes of experience that we enjoy when fantasising in dreams and in our imaginings. The difference lies in the absence of the generative process. This much is already implied by the notion of perceptual experience being characterised by its own unique temporal signature – one of the key components of the DEUTS argument. We have argued that active inference is quite consistent with this key tenet of the DEUTS argument. At least it is once one takes care to recognise how the generative model is actively maintained on the basis of a wide generative process that reaches all the way out into the environment.

In the next chapter, we turn to consider a different way in which to push-back against our argument for extended consciousness. The argument starts by using predictive processing to arrive at a criterion for demarcating the boundaries of the mind. This criterion is then used to argue for a global form of internalism, which seeks to undercut all arguments for the extended mind. We will show (perhaps predictably) that predictive processing implies no such thing while nevertheless retaining the criterion for demarcating the boundaries of the mind that this argument is premised on. We achieve this result by appealing to the second tenet of third-wave extended mind – the claim that the boundaries to the mind are open-ended and flexible.

Note

1 See Windt (2017) for detailed discussion of simulation views in dreaming.

4 Flexible and open-ended boundaries – Markov blankets of Markov blankets

4.1 Introduction

In this chapter, we begin by revisiting a recent objection that Chalmers developed against the DEUTS argument for extended consciousness (Chalmers forthcoming). The DEUTS argument is premised on the dynamic entanglement between internal and external processes, which arises out of tight couplings between perception and action. Chalmers argues that perception and action serve to delineate the boundaries of the mind. He tells us that

> [It] is natural to hold that perception is the interface where the world affects the mind, and that action is the interface where the mind affects the world. If so, it is tempting to hold that what precedes perception and what follows action is not truly mental.
>
> (2008, p. xi)

We will label this common-sense idea about how to demarcate the boundaries of the mind the "perception-action criterion." Chalmers (2008) argues that common-sense considerations like these present the strongest objection to the extended mind, and therefore to the extended conscious mind. Chalmers thinks that there is a way to save the extended mind but not extended consciousness. On broadly functionalist grounds, he recommends a revision of common sense for extended mind cases, such as the famous case of Otto. But Chalmers thinks the perception-action criterion undermines any argument for extended consciousness. As Chalmers sees it, if an external element – a notebook, say – is to be part of the realisation base for consciousness, then the information in the notebook must be directly available for global control. But this is not the case for information found in anything extra-neural, such as a notebook (as discussed earlier, in Chapter 2). When combined with the idea that perception and action delineate the boundary for the mind, Chalmers concludes that "extended consciousness is impossible" (forthcoming, p. 9).

Following a similar dialectical strategy to the previous chapter, we will show how Chalmers's common-sense criteria for demarcating the boundaries of the mind finds some support in recent work on predictive processing. Hohwy (2016, 2017a) formalises the perception-action criterion by appealing to the concept of a Markov blanket and approximate Bayesian inference.

The question we take up in this chapter is what the Markov blanket formalism implies for the debate over the boundaries of mind. Does it also provide support for the conclusion that minds do not and cannot extend into the world? We will begin by showing how Hohwy has used the Markov blanket formalism in just the same way as Chalmers uses the perception-action criterion to make an argument for internalism. We go on to challenge Hohwy's argument and by extension Chalmers's, given that both rely on the perception-action criterion. The focus in this chapter is thus broadly on the issue of how to delineate the boundary of the mind. However, insofar as the target of Chalmers's two-step argument is extended consciousness, the results of this chapter also bear on the more specific question of where to draw the boundaries of the conscious mind.

Our argument begins by discussing Hohwy's claim that the Markov blanket forms a boundary to the mind. We argue that Hohwy is wrong to presuppose that there is a singular unified and stable boundary that defines, once and for all, the boundaries of the mind. We argue, by contrast, that the boundaries of the mind are open to constant renegotiation. This implies that there is no unique and fixed way by which to demarcate the boundaries of the mind. We rely in part on Clark's (2017a) metamorphic argument for the idea that the Markov blanketed boundaries of an organism are flexible (see also Clark 2017b). However, we build on his argument by setting up our own view of nested and multiscale Markov blankets for the mind. We show how organisms are best understood as composed of *nested Markov blankets* (Kirchhoff et al. 2018; Palacios et al. 2017; Ramstead et al. 2017). The nesting of Markov blankets allows us to expand on what it might mean to dissolve cognitive agents into multiply nested dynamics of coordination and coalescence among neural, bodily, and environmental elements. We argue that the boundaries of mind are not permanently fixed but instead are plastic and continuously open to negotiation through action.

4.2 Hohwy on the Markov blankets of the mind

Hohwy (2016, 2017a) wants to establish that there is a sharp line to be drawn between internal mental states and external non-mental states, where external states include bodily and worldly states. His argument is elegant yet complicated, drawing on some technical tools from the machine learning and graph theoretic literature on causal networks to make his case. The

Markov blanket is at the core of his argument. The Markov blanket formalism provides a means of mapping Chalmers's common-sense intuitions about perception and action onto the causal dynamics of living systems (see Box 4.1 for technical details about Markov blankets).

The notion of a Markov blanket is taken from the literature on causal Bayesian networks (Pearl 1988; Hohwy 2016, 2017a). In living systems, the Markov blanket allows for a statistical partitioning of internal states (e.g., neuronal states) from external states (e.g., environmental states) via a third set of states: active and sensory states. This means that a Markov blanket both segregates internal and external states and couples them through active and sensory states. This formalism would seem to imply that perception and action indeed serve as interfaces connecting mind and world, as the perception-action criterion suggests. At least this much would follow if active states can be mapped onto action and sensory states onto perception. Internal states of a system cause active states, which influence, but are not themselves influenced by, external states. This can be read as the equivalent of the claim that action is the interface where the mind affects the world. External states of the environment cause sensory states that influence but are not influenced by internal states. This can be read as equivalent to the claim that perception is the interface where the world affects the mind. Thus, Hohwy concludes that "the mind begins where sensory input is delivered though exteroceptive, proprioceptive and interoceptive receptors and ends where proprioceptive predictions are delivered, mainly in the spinal cord" (2016, p. 276).

The Markov blanket forms what Hohwy describes as an "evidentiary boundary" for the mind such that anything located outside of the blanket has the status of a hidden cause that must be inferred before it can have any influence on behaviour. Hohwy argues that things "located in the environment cannot strictly speaking be said to be constitutive parts of our minds because things in the environment are outside [the Markov blanket], as are other people and their mental states" (Hohwy 2016, p. 269).

The first step of Hohwy's argument is that active and sensory states comprise a Markov blanket for internal brain states. Hohwy is not claiming that there is something essentially special about the wetware of the brain. He argues that internalism follows in virtue of the division between states induced by the Markov blanket, given that the system in question is engaging in approximate Bayesian inference. According to Hohwy, the observation that brain states engage in approximate Bayesian inference implies that those states are *self-evidencing*. Hohwy thinks of approximate Bayesian inference as a form of inference to the best explanation. In inference to the best explanation, when a prediction or hypothesis h_i best explains some occurring sensory evidence e_i, h_i provides evidence for itself.

Box 4.1 The Markov blanket and its partitioning rule

Pearl (1988) introduced the notion of the Markov blanket to denote a set of topological properties specific to Bayesian nets (Figure 4.1). The Markov blanket is the smallest set of nodes {2,3,4,6,7} that renders a target node {5} conditionally independent of all other nodes in the model {1}. The central point to note here is that the behaviour of {5} will be predictable by knowing the nodes making up its Markov blanket. This means that any node external to the system in question – in this case, node {1} – will be uninformative vis-à-vis predicting the behaviour of {5}. This means that once all the neighbouring variables for {5} are known, knowing the state of {1} provides no additional information about the state of {5}. It is this kind of statistical neighbourhood for {5} that is called a Markov blanket (Pearl 1988).

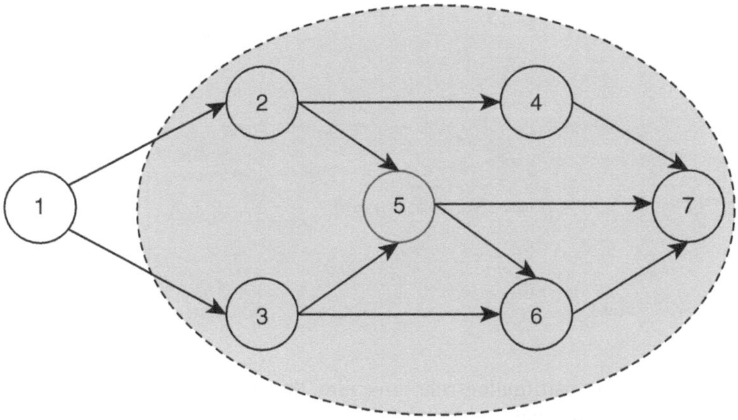

Figure 4.1 A schematic depiction of a Markov blanket with full conditionals
Source: Kirchhoff et al. 2018

Markov blankets enable us to speak about statistical dependence and independence among variables. Take a simple case of conditional independence. Observing it being cold could be explained by a window having been left open or by excessive air conditioning. Upon observing that the air conditioning is on full, the observation that it is cold now carries no information about whether the window is open (under the assumption that the two events are statistically independent). This example highlights that the coldness in the room and the window being open are conditionally independent, given that the air conditioning is on high. We can put this formally, noting that for any variable *A*, *A* is conditionally independent of *B*, given another variable, *C* if and only if the probability of *A* given *C* and the probability of *B* given *C* can be written as $p(A|C)$ and $p(B|C)$. In other words, *A* (it being cold) is conditionally independent of *B* (the window being open) given *C* (high air conditioning) if, when *C* is known, knowing *A* would provide no additional information about *B* (Beal 2003).

In the biological realm, Markov blankets are composed of active and sensory states (Friston 2013). It is the presence of active states and sensory states that separates internal states from external states, and vice versa. For example, the Markov blanket for a cell (the equivalent of node {5} in Figure 4.1) renders the internal states of the cell statistically independent from its surroundings, and vice versa (Figure 4.2).

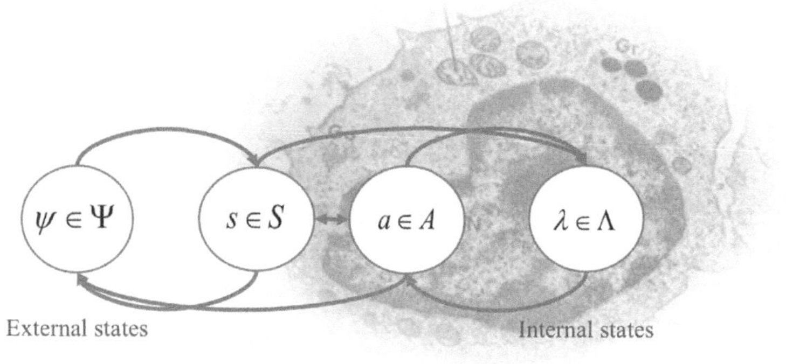

Figure 4.2 The partitioning rule governing the dependencies and independencies induced by a Markov blanket

Source: Friston 2013, used with permission from the Wellcome Trust

This figure highlights the partitioning rule governing Markov blankets ($S \times A$) implies a partitioning of states into external, sensory, active, and internal states. Hidden states (ψ) cause sensory states (S), which influence, but are not themselves influenced by, internal states (I), while internal states cause active states (A), which influence, but are not themselves influenced by, external states (Friston et al. 2014). Internal and external states can therefore be understood as influencing one another in a continuous and reciprocal fashion given how sensory states couple (i.e., relate) external states to internal states, while active states couple internal states to external states.

Interestingly, this is precisely the kind of reciprocal and recurrent influence between internal states and external states – via sensorimotor couplings – that lies at the base of how cognitive systems self-assemble and maintain their organisation over time and that one finds in many arguments for extended cognitive processes and/or extended cognitive systems.

Specifically, e_i becomes evidence for h_i conditioned on h_i explaining e_i. According to Hohwy, when "h_i is self-evidencing, there is an explanatory-evidentiary circle (EE-circle) where h_i explains e_i and e_i in turn is evidence for h_i" (2016, p. 264). The EE-circle is the Markov blanket. Hohwy takes self-evidencing processes to establish a Markov blanket for the mind, given that causes outside the evidentiary boundary need to be inferred by states internal to the boundary.

This suggests that approximate Bayesian inference is not an additional theoretical postulate over and above the Markov blanket formalism. It is rather a hypothesis about the Markov blanket formalism as it applies to the causal dynamics distinctive of biological agents (Friston 2013). We say more about this below in our exposition of Hohwy's argument.

What is distinctive about the internal organisation of biological agents is that they adapt to their environments by instantiating generative models that accurately and precisely predict their own sensory and active states. A central requirement for organisms that succeed in prediction error minimisation is that they do a good job of managing error on average. The reason for this is that the prediction error that the organism aims to keep to a minimum "is defined in terms of the states the creature tends to occupy in the long run (states which define its phenotype)" (Hohwy 2016, p. 270; c.f. Friston 2013).

According to Hohwy, it is the states of the central nervous system that an organism must maintain within viable bounds. Anything outside the nervous system is a hidden external cause about which predictions must be made (Hohwy 2016). Hohwy is led to claim on this basis that there is a "principled distinction between the internal, known causes as they are inferred by the model and the external, hidden causes on the other side of the Markov blanket" (2017a, p. 7). The Markov blanket formalism provides Hohwy with what "seems a clear way to define internalism as a view of the mind according to which perceptual and cognitive processing all happen within the internal model, or, equivalently, within the Markov blanket" (2017a, p. 7).

The next part of Hohwy's argument is to show how the presence of a Markov blanket constitutes a boundary between internal and external states, rendering these states *conditionally independent*. This is inherent in the first step. We add it as an additional part for purposes of exposition and clarity.

As we saw in Box 4.1, the idea of conditional independence can be defined in the following way: if a state *A* is statistically independent of a state *B* given *C*, where *C* denotes prior expectations about *A* and *B*, then, when *C* is known, knowing *A* carries no information about *B*.

It is relatively straightforward to connect this characterisation of statistical independence with the idea that only the brain functions to minimise prediction error. Specifically, once the Markov blanket for internal states is known, knowing the external (bodily and/or worldly) states carries no further information about the internal states. This implies that the internal states are in principle predictable from the states of their Markov blanket alone and are therefore statistically insulated from external states.

The final step of Hohwy's argument brings us back to consciousness. Optimising evidence for internal models is central in the account Hohwy gives of perceptual consciousness. The optimisation of evidence refers to the idea that any system that minimises long-run prediction error is at the same time maximising evidence for its own predictions about the most likely causes of sensory stimuli. This means that prediction error minimisation can be understood as optimising the "fit" between the causal structure a generative model maps and the actual causal and statistical regularities of the embedding environment. Hohwy hypothesises that conscious experience is associated with the posterior distribution that best accounts for sensory evidence over the hierarchy as a whole. As Hohwy puts it, "what gets selected for conscious perception is the hypothesis or model that, given the widest context, is currently most closely guided by the current precise prediction errors" (Hohwy 2012, p. 5). Conscious experience is thus modelled as the result of the brain's operations arriving at a particular internal model that best predicts the external causes impinging on the sensory states of the brain's Markov blanket.

To provide an illustration of what Hohwy has in mind, consider the following thought experiment. Hohwy (2013) invites you to imagine that you are inside a house. It has no windows, no doors, no Internet, and so on. The house provides you with no direct means by which to access its exterior. Suddenly you experience a tapping sound, and you wish to figure out the cause of this sound. Hohwy then asks you to imagine that you are a brain, that the walls of the house make up a Markov blanket separating what is inside the house from anything external to it, and that the tapping sound is your conscious experience (i.e., sensory observation). Note that there are all kinds of different plausible causes of your observation. It could be a woodpecker that is pecking away at the walls of your house. It could be kids knocking on the door or throwing stones. It could be something to do with your interior heating system (your homeostatic system), and so on. How would you then determine the cause of your observation?

Suppose now that the internal states of the Markov blanket–bounded system have the function of approximating Bayesian inference. The internal states are states of an internal generative model that accounts (approximately) for the sensory observations. The ultimate (Bayesian) goal is for internal states of the model to infer the probability of a prediction or hypothesis, *H* (e.g., it is a woodpecker at work), given the evidence, *E* (where *E* is the tapping sound you are hearing). In other words, the brain must do work to infer the posterior probability (or posterior density). The formal notation is written $P(H|E)$. On the assumption that the brain is an approximate Bayesian inference machine, it follows that the internal states comprising the brain can approximate the posterior density and thereby account probabilistically for the causes of your observations. Approximating the posterior density is possible, as the posterior density is equal to the conditional probability of the evidence, given a hypothesis, $P(E|H)$, multiplied by the prior probability of another hypothesis, $P(H)$, divided by the marginal likelihood of the evidence, $P(E)$. Thus, via approximate Bayesian inference, your internal brain states allow you to update the posterior under a generative model by following (on average) Bayes's rule (see Box 4.2 for technical details about Bayesian belief optimisation via Bayes's rule and the Kullback-Leibler divergence).

Hohwy's Bayesian account of consciousness postulates that the states inside a Markov blanket are involved in inferring a probability distribution containing any number of different hypotheses – e.g., that the causes of you hearing a tapping sound are likely be $H_1 \ldots H_n$, where these may be a woodpecker at work, kids throwing stones, or you hallucinating. It is the hypothesis with the highest posterior probability (i.e., with the least KL-divergence) that gets to determine the kind of conscious experience that you are having. This means that all of the work of realising the conscious experience you come to enjoy is done from inside of the brain, behind the Markov blanket separating the brain from the rest of the environment. Indeed what is true of consciousness is for Hohwy equally true of cognition more generally. Cognition in all of its forms is realised by processes that take place inside of the heads of individual, behind the blanket that bounds the mind.

4.3 Markov blankets: one or many

The Markov blanket formalism as applied to systems that approximate Bayesian inference serves as an attractive statistical framework for demarcating the boundaries of the mind.

Unlike other rival candidates for "marks of the cognitive" the Markov blanket formalism has the virtue of avoiding begging the question in the extended

Box 4.2 Bayes's rule and the Kullback-Leibler divergence

The standard notation for Bayes's rule follows:

$$P(H|E) = P(E|H)P(H)P(E)$$

The application of Bayes's rule is also known as Bayesian belief optimisation. The Kullback-Leibler divergence, or relative (Shannon) entropy, measures the success of Bayesian belief optimisation. We can write the Kullback-Leibler (KL) divergence as follows:

$$D_{KL}\,[P(H) \| P(H|E)].$$

The KL-divergence is a measure of prediction error (aka relative entropy). This means that it is a measure of information gained when accounting for the uncertainty between two probability densities, as the brain revises its predictions *from* the prior probability density, $P(H)$, *to* the posterior probability density, $P(H|E)$. Furthermore, the relative uncertainty specified by the KL-divergence is either higher than or equal to zero. The simplest and most intuitive way of illustrating this is that the KL-divergence is a measure of how different two probability densities are. This means that its minimum value should be at a point at which the two probability density distributions are equal – that is, that the difference between the two distributions is zero. The KL-divergence represents the *extra* number of samples (inferences) needed to account for the observation, as one is using the prior to approximate the posterior. The extra number of samples shows that the divergence is higher than zero, and that the divergence is equal to zero if and only if $P(H|E) = P(H)$. This is equivalent to the claim that the brain is in the game of inducing a probabilistic mapping from hidden causes to sensory observation under a model – a generative model. What is important here is that the smaller the divergence between the two densities, the better aligned the brain's predictions are with the true posterior (the actual cause of your sensory observation) – the better a grip it has on the dynamics of the environment.

mind debate. A number of proposals have been made for so-called "marks of the cognitive" – scientifically informed, but locationally uncommitted accounts of what it is for an element to count as a part of a cognitive state or process

(Wheeler 2010). These include, but are not limited to, the following two proposals. First, any state of a cognitive system must traffic in non-derived content or original intentionality (Adams and Aizawa 2008). This proposal has been argued to beg the question against the extended mind (Clark 2008; Menary 2007). Furthermore, there is currently no accepted naturalistic account of non-derived content (Hutto and Myin 2013). In the absence of such an account it is hard to know how to determine when a process meets this condition, and when it fails to do so. Second, it has been proposed that we look to the kinds of functional processes identified in our best theories of human cognitive psychology. These will tell us under what conditions an element counts as a part of a larger functional process that brings about some cognitive behaviour. But only including elements in the realm of the cognitive by dint of their human (or, better, brain-based) functional profile threatens to imply an unpalatable form of biochauvinism (Chalmers forthcoming; Sprevak 2009; Wheeler 2010).

The Markov blanket concept escapes these problems. It provides a statistical partitionioning of internal and external states given a third set of states – sensory and active states – rendering internal and external states conditionally independent of one another. While we dispute Hohwy's conclusion, we think his proposal to use Markov blankets may provide an elegant solution to the debate surrounding the boundaries of the mind. At the same time such a proposal faces an immediate difficulty as Hohwy himself recognises (2016; c.f. Anderson 2017). It threatens to lead either to a shrinkage of the mind, or to a proliferation of agents over time.

Hohwy says the question of where to place the boundary of the mind is easily settled in predictive processing models of the mind. The presence of a Markov blanket renders internal and external states conditionally independent of one another. Once the states of the Markov blanket are known in conjunction with the prior expectations of a system, causes outside the system will not be informative with respect to predicting the behaviour of that particular system. All other events must be inferred given the presence of a Markov blanket shielding internal states from external states. Yet once we say that the presence of a Markov blanket sets the boundary to the mind, it seems to inescapably follow that the sensory input and active output is restricted not only to the "input with the cause most proximal to the brain" but also to "the chemicals at the nearest synapse, or the ions at the last gate" (Anderson 2017, p. 4). As Anderson has noted, there "is always a cause even 'closer' to the brain than the world next to the retina or fingertips" (2017, p. 4).

The appeal to the presence of a Markov blanket as a principle for demarcating the boundaries of the mind thus threatens to lead to a shrinkage of the mind. Why privilege the sensory organs and argue that anything external to the sense organs has to be inferred? Why not instead take the Markov blanket to be formed out of sensory and active states internal to the brain higher up the neural hierarchy and treat the sense organs as external to the mind? We call this the shrinkage problem.

Hohwy tries to resolve the shrinkage problem by assigning an explanatory priority to the Markov blanket surrounding the nervous system, thereby underwriting its inferential seclusion from the rest of the body and world. Hohwy argues that this blanket is likely to be the explanatorily interesting boundary for predictive processing accounts of the mind (Hohwy 2016, pp. 273–274). The reason for this is that the agent needs to ensure that the states of the nervous system remain within a narrow, well-circumscribed range of values on average over time if it is to succeed in predicting its own sensory and active states. As Hohwy says,

> the agent worthy of explanatory focus is the system that in the long run is best at revisiting a limited (but not too small) set of states. It is most plausible to think that such a minimal entropy system is constituted by the nervous system of what we normally identify as a biological organism: shrinked agents are not able to actively visit enough states, and extended agents do not maintain low entropy in the long run.
>
> (2016, p. 274)

Is Hohwy right to conclude the central nervous system is always the best explanatory candidate for the boundary of the mind? Markov blankets can be defined at each of the multiple scales of organisation within a biological agent, including its cortical processing hierarchy. The statistical form of a Markov blanket will be the same all the way down and up – that is, all the way down to neurons and all the way up to embodied organisms (Allen and Friston 2016; Clark 2017a; Kirchhoff et al. 2018). What is important is that states internal to the boundaries of the system do a good job of predicting the sensory and active states of the system. But each layer in the neural hierarchy will be composed of internal states that predict the sensory and active states of the layer below. The input to each layer constitutes evidence for the model instantiated by this layer. Now, we have been arguing that Markov blankets identify the boundary of the agent. The agent can be identified with the states of the internal model it instantiates. Should we not conclude that each agent is composed of a nesting of smaller agents, much like a Russian nesting doll (Allen and Friston 2016; Clark 2017a)? Does each individual contain a multitude of smaller agents, one for each layer of its internal hierarchical model?

Hohwy (2016) considers this option. Instead of focusing simply on the Markov blanket that bounds the nervous system, he also suggests one could just bite the bullet and embrace the proliferation of agents. This would imply a view of organisms composed of many agents – that is, of many models that do work to garner evidence for their own existence. One might then "apply the contrastive method to establish which agent is the focus of explanation. One may ask 'why did this behavior rather than that behavior occur?' The agent responsible for the contrast is the agent of interest" (2016, p. 273). So we effectively have two options: either take the organism to be composed of a nesting of agents, or take there to be a single blanket of interest fixed in its location around the central nervous system.

We propose to use the Markov blanket formalism to delineate a single boundary for the mind while also holding that the mind is composed of a nesting of such blankets. At first glance, this may seem to generate an inconsistency. Doesn't our claim that Markov blankets can be used to define a boundary for the mind require us to privilege one of the many Markov blankets that the organism contains and treat this blanket as constituting the boundary to the mind?

We agree with Hohwy that the imperative to minimise long-run prediction error is what allows biological organisms to maintain a bounded organisation over time. So it is not the Markov blanket on its own that establishes the boundary of an agent but rather Markov blankets that are continuously generated through temporally extended processes of prediction error minimisation. This is a process that involves the whole nested, multiscale organisation of the organism in its dynamical coupling with its environment.

We dispute, however, that there is a genuine choice to be made between taking the organism to be bounded by a single Markov blanket and thinking of the organism as composed of multiple nested blankets. We are suggesting (and Hohwy would agree) that Markov blankets over time and within biological organisms generate processes that approximate Bayesian inference by minimising prediction errors in the long run. It follows that to identify the boundaries of the mind, it is the whole biological individual over time that we need to look for among this unfolding causal dynamics. The real question is therefore whether this biological individual must have a unique and fixed boundary or whether the same individual can be marked out by a boundary that shifts its location over time.

4.4 The metamorphosis argument

Must organisms have a single, fixed, and permanent Markov blanket around the nervous system, as Hohwy has argued? Clark (2017a) challenges such a conclusion by reflecting on the life of metamorphic insects. We call this the metamorphosis argument.

Clark invites us to consider the transformation of a caterpillar into a butterfly. As Clark (2017a) sees it, two important implications follow from looking at metamorphic insects. The first is that the actual generative model seeking to minimise prediction error or optimise Bayesian model evidence is the entire organism. The Markov blanket serves to individuate the organism from external states of its environment (Allen and Friston 2016). If Clark is right, then processes of self-evidencing are delegated not simply to the nervous system of an individual. The self-evidencing process involves a delicate spread of activity across states comprising the entire organism. In addition, it is not merely the model – that is, the agent or phenotype – at a specific temporal instant t that bounds prediction error but also the model as a temporally extended whole over the entire lifespan of an organism. The entire cycle of morphing from one phenotype – a caterpillar – into another phenotype – a butterfly – picks out the relevant Markov blanket for the caterpillar-cum-butterfly.

The second implication is that the organism itself should be thought of as an unfolding process of building and rebuilding the relevant boundaries that constitute the basis for minimising long-term prediction error. As Clark puts it, "Such agents 'knit their own Markov blankets' in ways that can change over time, without the agent thereby ceasing to exist" (2017a, p. 11). This picture of the agent, composed of metamorphic Markov blankets, gives us a way to conceive of the boundaries to the mind as comprising the entire organism. The whole organism is involved in the process of minimising prediction error, thus generating evidence for itself. It also provides a neat response to Anderson's shrinkage objection. The relevant place to look for the boundary to the mind is the shape-shifting Markov blanket of the whole embodied agent. There is thus no fixed and stable boundary (at least in metamorphic insects) that defines what is inside the agent and what is outside. Clark argues that what is true of metamorphic insects is also true more generally of extended minds. The boundary of the mind, he argues, is always up for negotiation. Cognitive systems are softly assembled out of elements that do the best job of efficiently and reliably accomplishing a task. Thus, in arguing for the flexibility and plasticity of the mind-world boundary, Clark is in total agreement with the third wave.

Interestingly, however, when it comes questions about the boundaries of the mind, Clark is sceptical about whether the Markov blanket formalism can help. He points out that Markov blankets are multiscale and nested:

> a single adaptive system will typically comprise multiple blanketed organisations – blankets of blankets of blankets. Which ones (if any) should count as tracking, at a given moment, the machinery of mind is precisely the question at issue. Such debates have (for better or for worse) been central to the philosophical and scientific debate concerning the extended mind . . . The question stands – and all the old

> arguments, pro and con apply – even if the neural contributions are all understood through the lens of Markov blankets, prediction and prediction error minimisation.
>
> (Clark 2017a, p. 8)

Clark concludes that Markov blankets are unfit for purpose when it comes to defining the boundaries of the mind. We agree with Clark that Markov blankets that combine to form an organism are nested and hierarchically organised (Figure 4.3).

The statistical structure of the Markov blanket is the same across all scales of an agent's hierarchical organisation such that this structure scales up and down recursively (Kirchhoff et al. 2018; Palacios et al. 2017). In the brain, for example, the functional interactions and cortical dependencies of

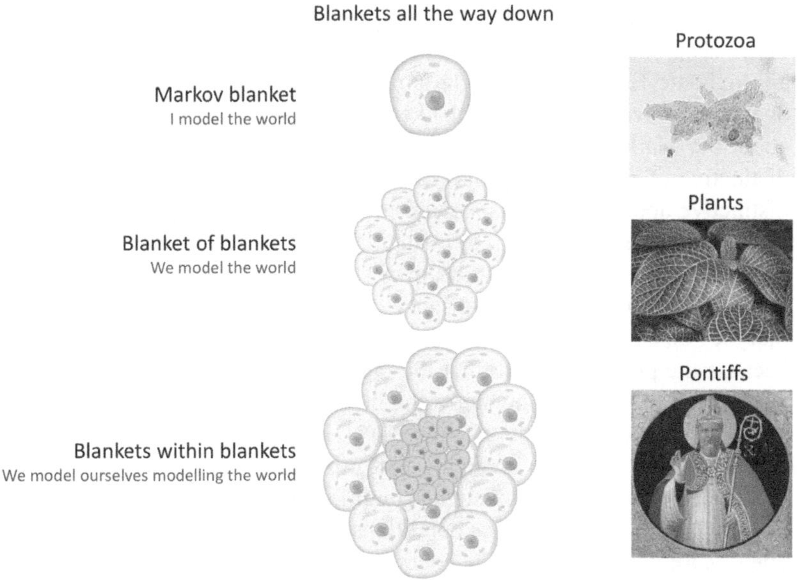

Figure 4.3 Schematic depiction of Markov blankets. The top figure depicts a single Markov blanket. The middle figure represents a multiscale and nested organisation of Markov blankets. The final figure suggests that cultural practices can envelop a multiplicity of individuals given its nested structure. Thus, Figure 4.3 represents the Markov blanket organisation all the way down to individual cells and all the way up to complex organisms like human beings.

Source: Kirchhoff et al. 2018

individual neurons, small neuronal ensembles, and even larger functional brain networks can be understood through the formalisms of Markov blankets (Pezzulo and Levin 2018). In the body, one finds nested hierarchies of cells forming organs and organs that make up whole organisms. It is even possible to extend this view of nested Markov blankets to elements of the environment. For example, human beings perform tasks while at work by collaborating with one another, thereby extending their Markov blanket beyond their own individual boundaries to include other people (Kirmayer 2018). We disagree, however, that this feature makes Markov blankets useless for determining what is and what is not a part of the boundaries of a system, for reasons we will explain in the next section.

4.5 Nested Markov blankets and the boundary of the mind

To see why Clark's worry does not undermine the use of Markov blankets to bound the mind, we need to briefly return to a point we made earlier in the chapter about self-evidencing. The hypothesis that biological agents approximate Bayesian inference is not an additional theoretical postulate over and above the Markov blanket formalism. It is rather a hypothesis about the Markov blanket formalism as it applies to the causal dynamics distinctive of biological agents (Friston 2013). It is proposed that what is distinctive about the internal organisation of biological agents is that they adapt to their environments by instantiating generative models that accurately and precisely predict their own sensory and active states. A central requirement for organisms that succeed in prediction error minimisation is that they do a good job of managing error on average and over the long run, because the prediction error that the organism aims to keep to a minimum "is defined in terms of the states the creature tends to occupy in the long run (states which define its phenotype)" (Hohwy 2016, p. 270; c.f. Friston 2013). This is the same as saying that minimising prediction error in the long run is to actively optimise the conditions required to sustain oneself in existence. It is to generate the conditions that allow for the preservation of one's Markov blanket over the course of a lifetime. An agent that succeeds in keeping its prediction errors to a minimum will have active and sensory states that are well predicted by the internal states of its model. It must sustain over time internal states that do a good job of predicting its own active and sensory states. Such a system will be "self-evidencing" – it will be gathering evidence in its sensory exchanges with the environment that maximise the probability of its own continued existence (Hohwy 2017a).

Markov blankets can thus bound the mind, their nested organisation notwithstanding. Markov blankets form over time based on conditional independencies between internal and external states. The internal states and the Markov blanket they entail determine what is and what is not part of the process of minimising prediction error. Anything that is not an internal state, an active state, or a sensory state cannot be part of the process of minimising prediction error and therefore cannot be a constituent part of a creature's mind (Hohwy 2016, p. 269). Minimising long-run prediction error is equivalent to generating the conditions that allow for the preservation of one's Markov blanket over the course of one's lifetime. We can thus use long-term prediction error minimisation to say what does or does not count as part of an individual's mind. This is because long-term prediction error minimisation establishes and maintains the Markov blankets that bound the mind, providing a partitioning of internal from external states.

This argument does not imply, however, that an individual's mind must have a single and fixed boundary over the course of its lifetime. What determines the location of the boundary are the sensory and active states that make up the blanket around the biological individual. These sensory and active states need not coincide in location with the central nervous system of an agent. To see why not, we will briefly provide two examples. First, consider the spider and its web. The best-known method by which spiders catch prey is via their self-woven, carefully placed, sticky web. The vibrations in the web can be thought of as forming sensory observations for the spider. The fly or whatever other creature is ensnared in the web is the external cause of the spider's sensory observations. The generative model that the spider embodies must produce a probabilistic specification of how sensory observations (outcomes) follow from a world of flies and other edible critters (external causes). This specifies the Markov blanket – the required conditional independencies between internal and external states that constitute the boundary of the prey-catching system. The web extends the sensory surfaces of the spider such that the Markov blanket has among its sensory states the vibration-sensitive web.

The web itself and the actions performed by the spider to construct it partially constitute the spider's sensory and active states. The internal states of the system comprising the spider plus its web forms the spider's generative model. It is the world of flies and other edible critters that are among the targets of prediction by internal states. Active states enable the spider to influence the sensory input it seeks out. The sensory states it samples in turn contribute to its maintaining itself in the bodily states it expects to occupy given its phenotype. From the partitioning rule governing the Markov blanket, it follows that the web is among the sensory states that, together with its active states, form a Markov blanket for the temporally and spatially extended spider-plus-web system.

Someone might object, however, that the spider catching its prey does not involve cognition in any interesting sense. This example thus does not carry much weight when it comes to making an argument from predictive processing for the extended mind. Consider as a second case the by now overly familiar case of Otto and his beliefs (Clark and Chalmers 1998). When Otto forms the intention to visit Modern Museum of Art (MoMa), he predicts the sensory states associated with the action of heading off to the museum. He then acts in such a way as to fulfil his predictions, thereby reducing prediction error. Among the actions Otto must perform to fulfil his predictions is the consultation of his notebook. Were he not to consult his notebook, he would remain in a state of high prediction error, since he would not be able to make his sensory observations fit with his prior expectations. Otto's notebook, we suggest, is among the active states – the generative process that forms the boundary to Otto's mind. The external causes of his sensory states originate in the world that Otto is trying to navigate to get to the museum. They include things like the world of traffic lights, buildings, pedestrians, cars, and so on – in short, all of the obstacles that stand between Otto's current location and the museum. They are what Otto is uncertain about, and his active manipulation of his notebook is a means to reducing this uncertainty. We suggest, then, that the self-evidencing condition should likewise push us towards acknowledging the contribution that the notebook makes to long-term prediction error minimisation for Otto.[1] The active states of the Markov blanket in Otto's case will include his states of manipulating the notebook. We therefore conclude that active states – Otto using his notebook – can be part of the states minimising prediction error and so can be part of the states comprising the boundary to the mind.

Given the nesting of Markov blankets, how do we avoid the two horns of the dilemma we outlined in Section 4.2? How do we avoid, on the one hand, the shrinkage of the mind and, on the other the proliferation of agents, one for each of the Markov blankets that make up the agent's mind?

The statistical structure of the Markov blanket allows for the formation of Markov blankets at larger and larger scales (of cells, of organs, of agents, of local environments). The reason is that the organisation of Markov blankets occurs recursively. This means that Markov blankets at larger scales – e.g., the spider and its web – have the same statistical form as the Markov blankets found at smaller microscopic scales – e.g., the Markov blanket of a single cell comprising a spider. At any level of hierarchical organisation, we find a Markov blanket consisting of active and sensory states that separates the internal states of this layer of the generative model from the external states it tries to predict (Figure 4.4). The organism and its niche will be "defined not by a singular Markov blanket but by a near-infinite regress of

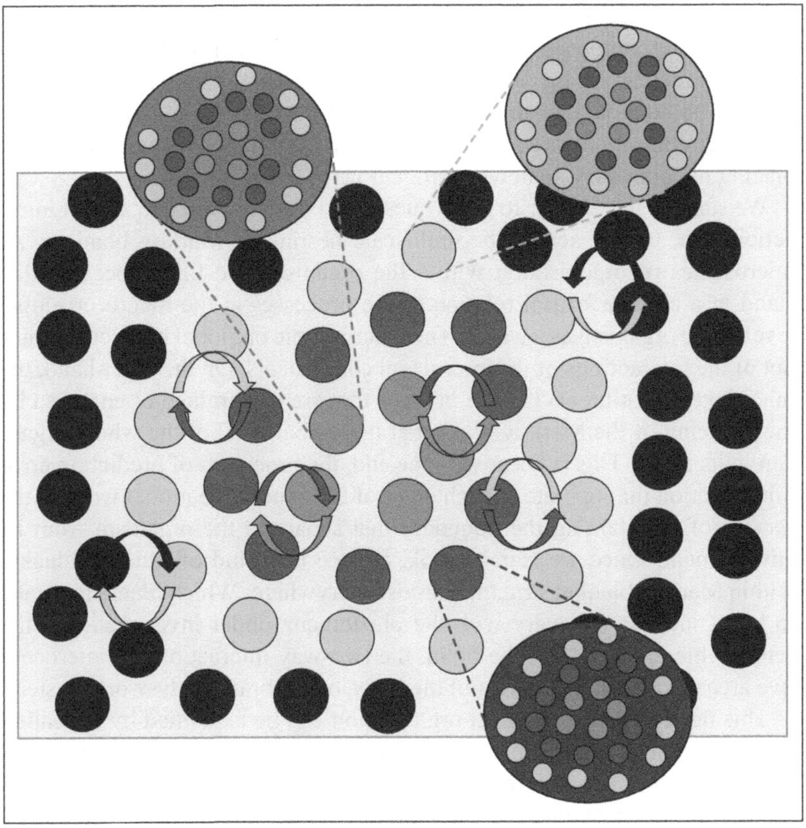

Figure 4.4 Markov blankets *of* Markov blankets. This illustrates the scale-free and nested organisation of Markov blankets, where each node in a global Markov blanket comprises it's own Markov blanket, and so on, at multiple temporal and spatial scales of systemic organisation.

Source: Kirchhoff et al. 2018; Palacios et al. 2017

causally interacting Markov blankets within Markov blankets" (Allen and Friston 2016, p. 19; Kirchhoff et al. 2018; Kirchhoff 2018; Ramstead et al. 2017).

The brain does not comprise a single Markov blanket. The organism is constituted by numerous interconnected and coupled Markov blankets. Indeed, the brain and body are composed of nested and multilayered Markov blankets, each of which is formed and maintained in generative processes (see Figure 3.1 in Chapter 3). The Markov blanket that marks the boundary of the mind distinguishing what is internal to the mind from what is external is, we suggest, the blanket found at the global scale of the whole organism situated in its local environment. The organism and its brain, however, are composed of a nesting of blankets, each of which induces a boundary distinguishing internal from external states. What is our argument for placing the boundary of the mind around the whole organism rather than thinking of the agent as a Russian doll composed of many smaller agents or placing the boundary somewhere higher up in the hierarchy?

We suggest the answer to these questions lies in the circular causal interactions that unfold across the multiscale nesting of Markov blankets. At microscales of organisation within the organism, we find processes that stand in a circular causal relation to the processes at the macro or global level of the organism as a whole. The macroscopic or global blanket emerges out of the interactions of the microlevel constituents. On the other hand, the microlevel constituents behave in ways that are constrained or enslaved by the inducing of the Markov blanket at the global level of the whole organism in its niche. This is because in the end, the processes of prediction error minimisation that operate at each layer of the generative model work in the service of maintaining the boundary that separates the organism from its environment. Once we start to look, we see this kind of Markov blanket within Markov blanket structure almost everywhere. Which blanket we take to be of interest will vary with the phenomena under investigation – the sensory hierarchy within the brain, the two-way interactions of interoceptive areas of the brain and rest of the body, or the brain-body-world system.

This nested Markov blanket organisation can be explained by appealing to the concept of an order parameter, also known as a collective variable. An order parameter is a macroscopic (global) feature of a system. It captures the dependencies among the parts making it up. When the statistical form of each constituent is explored, this means that each part can be associated with being an internal state of a larger Markov blanket, which in turn allows each internal state to influence and be influenced by all other internal states. This process of individual Markov blankets organising into larger Markov blankets reconfigures the dependence structure between internal, external, active, and sensory states at the scale below – as depicted in Figure 4.4. This kind of self-organising activity implies a separation of the dynamics involved into slow and fast timescales – a signature feature of the slaving principle in synergetics (Haken 1983). This makes it possible to understand how slow macroscale dynamics are determined by fast microscale dynamics. Since the macroscale Markov blanket acts as an order parameter, it follows that all the dynamics at the microscale no longer behaves independently of the macroscale dynamics but are instead, as Kelso puts it, "sucked into an ordered coordinated pattern" (1995, p. 8). The dynamics at the microscale are therefore constrained by the dynamics at the macroscale.

We can think of no better way by which to push back on the shrinkage problem. Recall the problem for both Chalmers and Hohwy is that once they insist on the boundary of mind as induced by perception and action, it is difficult to see how one could avoid the shrinkage problem. After all, the cell has a Markov blanket constituted by active and sensory states (Friston 2013). But the nested Markov blanket view escapes this threat. The self-evidencing dynamics exhibited at the macroscale, *n2*, constrains the self-evidencing dynamics at the microscale below, *n1*. Hence, it is the self-evidencing dynamics of the whole that determines the self-evidencing dynamics at smaller and smaller scales. Only when the dynamics of the macroscale ensemble feed back into the dynamics at the microscale, thus forming a prediction error minimising whole, can one sensibly talk about dynamics of Markov blankets at the macroscale.

Finally, the nested Markov blanket organisation does not imply a boundary between the organism and its environment that is fixed and non-negotiable, as Hohwy claims. The mind is bounded, but its boundary is best seen as "hard-won and fragile developmental . . . achievements, always open to renegotiation" (Sutton 2010, p. 213).

4.6 Summary

In this chapter, we have shown how to make use of the Markov blanket to identify the boundary of the mind. The Markov blanket has been used by Hohwy to defend a strongly internalist view of the mind. We have argued, however, that this concept has no such implication. It can be used as a criterion for demarcating the boundaries of the mind. But it does not have the implication that there is a single, fixed boundary separating the organism from its environment. On the contrary, we have shown that there is a flexible boundary that forms as the organism engages in actions that are aimed at ensuring that it remains in the sensory states it expects to occupy. In summary, there are no fixed, permanent, and non-negotiable boundaries to the mind.

Note

1 Clark argues for a similar conclusion – see Clark (2017a, §4). One objection to this line of argument might be that the contribution of the external resource to prediction error minimisation is too temporary and short term for it to meet our second condition. But this objection overlooks the work of Otto's notebook and his active use of it as part of the generative process for an agent like Otto, who can no longer rely on his biological memory.

5 Expectation and experience

A role for cultural practice in precision estimation

5.1 Introduction

The aim of this chapter is to understand the relation between expectation and experience. We take up two challenges from predictive processing that aim to show that sensorimotor enactivism implies an internalist, brain-bound theory of the realisers of phenomenal consciousness. The first argument targets a weak spot of sensorimotor enactivism already identified by those sympathetic to its externalist ambitions. The weak spot is the role of expectations in making possible experiences of the world (Hutto 2005; Rowlands 2007; Roberts 2010).

Sensorimotor enactivists argue that phenomenal experience of the world is mediated by sensorimotor understanding. But this causes difficulties for any argument from the sensorimotor theory to extended consciousness. It can now be argued that sensorimotor understanding is locally realised in the brain of an individual. Cognitive neuroscientist Anil Seth has made an argument along these lines (Seth 2014). He uses predictive processing to develop an explanation of sensorimotor understanding as realised in the generative model. Seth takes predictive processing to entail an account of phenomenal consciousness as intrinsically and locally realised in neural processes. We agree with Seth that predictive processing can provide a framework for understanding the role of expectation in shaping conscious experience. We disagree with him, however, that sensorimotor understanding plays a mediating role in the agent's active engagement with the world (Hutto 2005). We deploy the distinction introduced in Chapter 3, between the generative model and process, to argue that active engagement with the environment makes possible and orchestrates the expectations involved in counterfactual prediction.

Sensorimotor understanding underwrites one form of expectation that is important for making possible experience of the world. Sensorimotor enactivism, as we are developing the view, holds that there are many other forms of expectation that also play important roles in shaping what we experience.

This raises the question of how top-down expectations get to influence the brain's processing of sensory signals from the world in such a way as to partly constitute what a person consciously experiences.

In predictive processing, this question is answered by appealing to attentional processing understood in terms of precision estimation (Clark 2016; c.f. Feldman and Friston 2010). Informally, precision refers to the uncertainty associated with a prediction error signal in the neural hierarchy. At the same time as the brain is calculating prediction errors, it is also estimating the reliability and salience of those prediction errors. To give an example of one way that this can play out, suppose that prediction errors at a given level in the hierarchy are weighted as unreliable. This allows for expectations that are found at this level of the hierarchy to strongly influence processing – to override prediction errors. The influence of expectation on experience will depend on a neurally realised precision mechanism. Thus, a second internalist argument can be made based on the role of the neurally realised precision mechanism in determining the degree of influence of top-down expectations on conscious experience.

We respond to this objection by showing how precision can be weighted based in part on patterns of cultural practice. Patterns of activity found in a practice can lead to trust and confidence being assigned to priors that operate at high levels of processing in the sensory hierarchy. Assigning high precision to expectations at these higher levels of sensory processing leads to an alignment of expectations among people taking part in a practice. We will show how alignment has as a consequence a literal shaping of experience by a person's cultural niche.

This second stage of our argument thus amounts to an argument from distributed assembly. We aim to show that the prediction error minimising machinery is not always and necessarily wholly assembled by processes that take place within the individual. The precision mechanism that does the work of settling the relative influence of bottom-up and top-down flows of information within the brain can be tuned from outside of the individual. There are constraints in operation in the practice that can be viewed as playing an active role in assembling the prediction error minimising machinery within individuals. This is the argument from distributed cognitive assembly – a key tenet of the third wave.

5.2 The predictive processing account of sensorimotor understanding

Sensorimotor enactivists take experience to make available to us more of the world than is immediately present to us at any given time. Imagine you are holding a mug of coffee in your hand. What you see of the mug is its facing side. You feel the handle because this is what your hand is currently in contact with. Yet what your experience makes available to you is the whole mug of steaming coffee. Your experience goes beyond what you can immediately sense. It opens or reveals the world to you. This presence of the objects of experience is explained by enactivists in terms of expectations. You do not represent all of the hidden unseen detail in a picture-like image in your head. The hidden unseen detail is something you know how to access through movements of your head and eyes and through physically manipulating the mug, turning it about with your hands. Nor do you need to actually move in order to experience the presence or availability of the whole mug in your hand. All that is necessary for the whole mug to be present to you in your experience is that you understand what *would* happen *were* you to move or interact in some other way with the mug. This understanding or know-how consists of expectations more or less indeterminate, of what would come into view, what would show up of the mug were you to vary your relation to it. In the absence of these expectations, your experience would not reveal or make available to you the world in all of its phenomenologically textured richness. To experience the world as having perceptual presence thus requires understanding sensorimotor contingencies (SMCs).

Seth has proposed a predictive model of how the brain might "encode" mastery of SMCs (Seth 2014). He calls this theory the predictive perception of sensorimotor contingencies theory (henceforth PPSMC). Seth argues that a key selling point of PPSMC is that it can explain what happens on the side of the perceiver when they exercise mastery of SMCs. We can think of this in terms of active inference. Recall that in active inference, the brain predicts the proprioceptive and exteroceptive sensory input associated with a course of action. The result is prediction error because the current states of the body do not match those the subject would occupy were they to act. What is important for our current discussion is that these predictions can be thought of as reflecting the subject's understanding of SMCs. This is something that Seth claims remains a mystery in standard treatments of sensorimotor enactivism (e.g., O'Regan and Noë 2001; c.f. Roberts 2010).

PPSMC puts pressure on any inference from sensorimotor enactivism to the thesis of extended consciousness. Seth articulates the problem well:

> An interesting implication of PP [predictive processing] in this context is that it captures the close coupling of sensation and action emphasised by enactive approaches such as sensorimotor theory but without endorsing the strongly enactivist [or "extended mind"] view that the local material vehicles [the "minimal supervenience base"] for perception or consciousness could extend beyond the boundaries of the skull.
>
> (2014, p. 7, our additions in brackets)

Once we think of sensorimotor understanding as represented in the generative model, it can then be argued that sensorimotor enactivism entails an internalist explanation of phenomenal consciousness. What does the work of explaining sensorimotor understanding is the prediction error minimising machinery housed within the brain of the individual, not the active exploration of the environment as was argued in Chapter 2.

Seth's argument assumes, however, that counterfactual prediction can function independently of actively exploring the environment. We argued against this assumption in Chapter 3, where we showed how the generative model and the generative process stand in a relation of reciprocal dependence. The generative process is a real-world process that as the agent acts generates sensory observations. It relates external causes and sensory input to embodied actions. It is the generative process that serves to both orchestrate and maintain the distribution of prior expectations of the generative model. The generative process brings about the actions of the agent that control the sensory input to the brain. This gives action an ongoing role to play in enabling the instantiation of the generative model. Indeed, the generative model can be thought of as a consequence of the embodied modes of sensorimotor engagement brought about by the generative process. Specific patterns of sensorimotor engagement with the environment are inferred or predicted by the generative model.

In our view, perceptual presence is brought about through the coupling of the generative process to the generative model. Perceptual presence recall refers to the phenomenological feature of experiences of the world: they present the world in rich phenomenological detail that goes beyond what is currently being sensed by the subject at any given moment in time. Recall the example of your experience of the presence of the whole coffee mug when, strictly speaking, all you currently sense is its facing side. Perceptual presence is not just the result of the counterfactual predictions of the generative model. Also necessary is the actual bringing into focus of things in the environment through sensorimotor engagement with them.

It might be objected that this account of perceptual presence fails to explain sensorimotor mastery. As I am sitting here typing, I can right now experience the perceptual presence of my coffee mug without doing a thing or making a single movement. I see the presence of the whole mug in part because of my sensorimotor *expectations*. I expect that were I to change my relation to the mug, my movements would bring the currently hidden parts of the mug into view. Expectations play an important and necessary role in an account of perceptual presence. But they are counterfactual – they relate to the sensory effects of possible but non-actual movements.

Like Seth, we understand sensorimotor expectations in terms of the counterfactual predictions of the generative model. We are not claiming that the generative process is sufficient to generate conscious experience. Conscious experience requires the coupling of the generative model to the generative process. However, we depart from Seth in denying that sensorimotor expectation is sufficient for perceptual presence. Such expectations must be brought into contact with the actions that the generative process undertakes to yield perceptual presence. The world seems to be present to us in experience because experience is the result of active sensorimotor engagement with the world.

Clark (2018) asks why it is that experience presents us with a single determinate, unified, and coherent perceptual take on the world. Why do we not experience more uncertainty, ambiguity, and indeterminacy about how things are if perceptual experience is the brain's best "multilevel" guess about how things are in the environment? Clark suggests that perceptual determinacy and perceptual coherence are due to the need of the brain to select a single "best-fit model for the control of action and choice" (2018, p. 78). Thus, the role of the generative model in driving action accounts for why experience should present us with a unified and coherent interpretation of how things are. Although we agree with Clark that predictive processing serves the purpose of driving action, we do not agree with Clark's solution to his puzzle. Like Seth, Clark fails to take into account the distinction between the generative process and the generative model. We agree with Clark that action accounts for the unity and coherence found in perceptual experience, but the generative process brings about active sensorimotor engagement with the world. It is because we are actively engaged with a world that is for the most part coherent, unified, and determinate that our experience has these phenomenological features. So instead of locating determinacy in the single, coherent hypothesis that is selected to drive action, we would locate determinacy in the environment we engage with in action. The generative process and the role it plays in coupling the agent to a more or less determinate environment lend determinacy to our experience.

We turn next to the wider question of the role of expectation in shaping the phenomenal character of conscious experience. So far, we have responded to an argument that sensorimotor enactivism when combined with predictive processing implies a view of the realisers of phenomenal consciousness as local to the individual. In the next section, we present a second line of argumentation that attempts to derive internalism from predictive processing. This argument takes as its starting point research that shows the necessary part that expectation plays in generating what a person experiences.

5.3 Seeing what you expect to see

A claim common to both sensorimotor theories of consciousness and predictive processing is that expectation plays a profound and important role in shaping what a subject experiences. Consider, for instance, that the physical manipulations of a physiotherapist feel different when you are expecting them compared to when they take you by surprise. Following a description of a number of cases like this one, Noë (2016) has the following to say about bodily sensations:

> What these examples suggest is it that maybe we can't make a sharp line between a bodily sensation, and our thoughts, attitudes, beliefs and expectations about it, as well as our actions in response to it. It isn't that there is no difference between pains and other feelings, on the one hand, and what we say and think and do about them, on the other. Of course there is! It's that it is the very nature of pain, and other sensations, to be affected by and sensitive to what we say, think, know and expect, and so on, and importantly to be affected by these in characteristic ways.
>
> (p. 72)

Sensorimotor enactivism claims that as subjects of experience, we understand how what we experience varies in different contexts. Discriminating the colour of an object, for instance, rests on abilities for tracking objects with respect to colour under variations in illumination, across contrasting viewing conditions, as the object moves or with one's own movements. One expects a whole suite of variations in how the object shows up under these different contexts. In Noë's version of sensorimotor enactivism, perceptual experience is a "kind of thoughtful exploration of the world, and thought is, at least in a wide range of cases. . . a kind of extended perception" (Noë 2012, p. 45). Thought and perception to the extent that they can be distinguished are different modes of exploring and achieving access to the world.

The claim that the line between perception and thought is fuzzy is also central in predictive processing, as Clark writes:

> To perceive the world just is to use what you know to explain the sensory signal across multiple spatial and temporal scales. The process of perception is thus inseparable from rational (broadly Bayesian) processes of belief fixation, and context (top-down) effects are felt at every intermediate level of processing.
>
> (2013, p. 10)

Clark is arguing here that the line separating top-down (higher-level) beliefs or expectations from the bottom-up (lower-level) perceptual processing of sensory signals (i.e., sensory prediction errors) is not sharp. Perceptual systems make use of whatever combination of bottom-up sensory signals and top-down knowledge does the best job of minimising overall or global prediction error.

Thus, consider the following illustrative example from Lupyan (2015). You pour yourself a glass of orange juice in the morning, and unbeknownst to you, a prankster swaps the juice for milk. What happens when you take a gulp from the glass with the expectation of drinking juice? Do you taste milk or something that combines the smooth texture of milk with the sweet taste of orange juice? Lupyan cites evidence that the mere expectation of taste is sufficient to activate areas of gustatory cortex specific to the expected taste (Fontanini and Katz 2008, cited by Lupyan 2015, p. 13). Incoming sensory signals always occur in the context of top-down expectations, and whatever you end up experiencing will be the result of this context-sensitive processing of sensory signals.[1]

A similar effect is described by Clark (2016), this time for audition. Clark cites a study in which subjects were told if they listened closely to the sound file, they would hear a faint version of Bing Crosby's "White Christmas" among the white noise (Merckelbach and Van de Ven 2001; see Clark 2016, p. 54). In fact, the sound file contained only white noise. However, almost one third of the participants reported nevertheless hearing "White Christmas." It seems the strong expectation was sufficient to induce a mild auditory hallucination in these subjects. The top-down expectation was guiding visual processing in such a way as to lead the subjects to find a signal among the noise where there was no signal.

Now it might be objected that if belief and expectation can change the contents of experience, why then do illusions persist in the face of contradictory knowledge and beliefs?[2] Why is it that one continues to be susceptible to the Muller-Lyer illusion, for instance, even after one has discovered or been told that the lines are of equal length? Why does one's true belief that the lines are equal in length not correct one's perception of the lines being unequal in length?

The answer to this objection lies in the notion of precision – one of the key players in the predictive processing theory. The core claim of predictive processing is that processing over multiple levels of the sensory hierarchy combines prediction based on top-down priors and expectations with bottom-up sensory information. The key challenge that the brain always faces is thus how to balance the influence of prediction error at each layer of processing with top-down knowledge. This challenge is settled by the precision assigned to a prediction error.

Precision refers to the inverse variance of a prediction error – the estimated uncertainty or reliability associated with a prediction error relative to an organism's predictions. The reliability of prediction error is something that varies across different contexts. Thus, in addition to learning how to predict its own sensory input, the brain is also learning about the fluctuations in reliability of prediction error in different processing contexts. Precision is the brain's guess about how much trust to place in a prediction error signal.

The more precision a prediction error is expected to have, the greater the certainty that the prediction error is a signal and not mere noise. Prediction errors that are estimated to have high precision get to exert a greater influence on processing higher up in the hierarchy. This is sometimes described in terms of the "post-synaptic gain" on the error units being increased or "turned up" (Friston 2005). Prediction errors are assigned low precision when they are unreliable and thus get to exert a correspondingly weaker influence on downstream processing. When prediction error is weighted as imprecise, the "gain" is turned down on the error units, which has the effect of suppressing the error signal. "Attention" in predictive processing refers to the process by which gain is increased for prediction errors deemed most likely to carry important information because they bear on a current task or convey some risks or opportunities (Feldman and Friston 2010).

Precision corresponds with learning rate. If a person is certain that a prior is correct, they should be less willing to update this prior when they come across conflicting evidence. They already know what they need to know, so there will be little of interest to learn from new information. In this case, the learning rate with respect to this prior will be low, and they will assign low precision to evidence that comes into conflict with the prior. Conversely, learning rate is high when evidence is expected to have high precision. The organism then pays closer attention to the evidence because it carries potentially important information.

Now that we have the idea of precision in play, let us return to the question about why perceptual illusions are not updated or corrected once a person acquires a true belief. Perceptual illusions can be thought of as cases in which there is low-learning rate, which is to say that evidence is assigned low precision relative to the predictions of high-level priors. Thus, in the hollow-mask illusion, we experience the nose of the hollow-mask as concave and not convex, even after one has reached inside of the hollow-mask and felt that the nose points inwards and not outwards. This illusion arises because of a strong expectation that faces have noses that point outwards. Any evidence to the contrary will thus be down-weighted relative to the well-supported expectation that the nose one is currently seeing is concave and not convex. Whatever uncertainty there is on the matter is discounted. We see what we expect to see – namely a mask with a nose pointing out.

If the brain is to succeed in minimising prediction error over the long run, it must find the right balance between the accuracy and simplicity of the generative model. The brain should not update its priors in the case of perceptual illusion, since to do so would result in a model that is overfitting. It reduces error in the short term but ends up generating more error in the long run by, for instance, revising a well-supported generalisation that the noses on people's faces tend to be concave. The brain should equally avoid underfitting its model to its data, ignoring or down-weighting evidence that would otherwise allow it to identify important patterns. To process information optimally, the brain must estimate precision in such a way as to minimise *expected* prediction error – that is, in such a way as to keep long-term prediction error to a minimum (Friston et al. 2015a; Hohwy 2017b; Kiverstein et al. 2017).

We are now in a position to pose the second challenge of the chapter to the argument from sensorimotor enactivism to extended consciousness. We began this section by noting that a claim common to both sensorimotor enactivism and predictive processing is that expectation plays a vital role in shaping what people experience. We have just argued that the impact of top-down expectation on the perceptual processing that leads to conscious experience is settled by precision weighting. The precision of prediction error is weighted based on learning rate. Precision weighting, however, might be naturally thought to happen locally to the individual. Postsynaptic gain or precision have been hypothesised to be modulated in part by the neurotransmitter dopamine produced in the midbrain and projecting to the cortex via the basal ganglia. Dopaminergic response signals a change in the expected precision or uncertainty associated with a prediction error, which can shift learning rate by increasing the synaptic gain on error units sustaining the processing of an error signal (Feldman and Friston 2010; Friston et al. 2012a). But if this hypothesis is along the right lines, the explanation of how expectations influence the construction of experience will be one that needs to be told from within the individual. Thus, far from providing support for extended consciousness, sensorimotor enactivism seen through the lens of predictive processing actually supports internalism about the realisers of phenomenal consciousness.

In the remainder of the chapter, we develop a response to this second challenge by invoking a further key tenet of the third wave: the argument from the distributed assembly of cognitive systems. The precision mechanism as we have described it in this section can be thought of as doing the work of assembling the cognitive system that generates conscious experience. In the next section (Section 5.3) we show how the notion of cognitive assembly may apply to consciousness. The final section of the chapter (Section 5.4) shows how precision expectations are sometimes set on the basis of patterns of social and cultural practice. When this happens, the context-sensitive and task-dependent assembly of the cognitive system that leads to conscious experience can be thought of as distributed, or so we shall argue.

Thus, the role of precision expectations in balancing the relative influence of the incoming sensory signal and top-down expectation does not necessarily imply an organism-centred or organism-bound theory of cognitive assembly.

5.4 The cognitive assembly of consciousness

In Chapter 1, we explained that a key tenet of third-wave extended mind is that cognitive systems are softly assembled out of a variety of neural, bodily, and environmental components. The notion of soft assembly is typically used to draw a contrast between cognitive systems and systems like computers or cars, in which the component parts out of which these systems are built have fixed and dedicated functions (Anderson et al. 2012). Softly assembled systems, by contrast, are made up of a temporary coalition of elements. The functional contributions of each of the elements is set by the task for which the cognitive system is assembled. The cognitive system will be made up of whatever mixture of elements is able to accomplish the task most efficiently. Sometimes this will mean relying purely on internal resources. On other occasions it will mean making use of the right mixture of internal neural elements and bodily action on resources located in the environment (Clark 2008). It follows from this view of cognitive systems as softly assembled that the cognitive system has no fixed boundaries. The boundaries of the cognitive system form in action, as was argued in Chapter 4.

In predictive processing, it is the precision mechanism that settles in a particular setting, the coalition of processing resources that will best get the job at hand done. "Efficiency" in predictive processing is a property of the generative model. It is a feature of a generative model whereby it relies on only those parameters that are needed in the long run to keep prediction error to a minimum (i.e., those that avoid either underfitting or overfitting the agent's sensory observations). The generative model serves the generative process, or so we have argued. Thus, the parameters of the generative model must relate in some way to the organism's effective practical engagement with its environment (Bruineberg et al. 2016; Clark 2016, 2017b). A good generative model is not necessarily one that accurately mirrors reality but one that can facilitate successful action. This means that the parameters of the generative model will often include variables whose values are stable and regular features of the environment that are most likely to keep prediction error low when engaging with them. Clark gives the nice example of white lines painted on a winding, narrow road. The white lines allow the brain to avoid the otherwise computational costly task of computing the precise combination of twists and turns needed to stay on the road. Instead, all the brain needs to do is weight as precise prediction errors that relate to keeping the white line in the centre of the visual field.

The context-sensitive adjustment of precision expectations thus does the job of forming the exact coalition of elements needed to maintain the cycles of perception and action required for the performance of a given task. If this job of updating precision estimations is done internal to the individual, it follows that cognitive systems are organism centred even if they are not organism bound (Clark 2008). In the case of phenomenal consciousness, there are reasons to think that the cognitive system is not only organism centred but also organism bound, as we now explain.

In a recent review of research on the neural correlates of consciousness, Victor Lamme has argued that if anything is a defining feature of conscious processing, it is perceptual organisation (Lamme 2015). Lamme writes that the "grouping of image elements according to Gestalt laws and figure-ground segregation. . . depend strongly on the conscious state and on conscious perception" (2015, p. 21). He goes on to write that "if we want to identify visual functions that mark the transition from unconscious processing to conscious vision, grouping according to Gestalt laws (incremental grouping) and figure-ground segregation (or perceptual organisation in general) are our best bets" (ibid.). An example we will return to is the figure-ground segregation found in the Rubin vase illusion. We can see the vase as the figure and the two faces looking at each other as the background to the image. But we can also easily switch so that now the two faces become the figure and the vase moves into the background (Figure 5.1).

Lamme suggests that perceptual organisation might be associated with conscious processing because of the spatio-temporally extended neural processing that perceptual organisation seems to require. Diverse forms of information are

Figure 5.1 Rubin's vase illusion – depending on figure-ground segregation, we can switch between seeing the image as a black vase against a white background and seeing it as depicting a face-to-face encounter against a black background.

Source: Rubin, E. (1921). *Visuell Wahrgenommene Figuren*. Copenhagen: Gyldendal

brought together and combined to form coherent, unitary, meaningful experiences. It takes time for information to be combined from diverse regions of the brain into an organised percept. Lamme gives as an example seeing the face of a familiar person – let us suppose it is a childhood friend we have not seen for many years. We do not just see a face but the face of someone we have known and shared many experiences with in the past. Lamme suggests that our visual experience integrates "all possible information about that face" (2015, p. 25). This integration will depend on spatial and temporally extended neural processing involving many incremental steps and horizontal and feedback connections among different layers of processing.

The cognitive assembly of the cognitive system that results in conscious experiences can thus be thought of as the process of integration and organisation of processing elements, the result of which is a coherent, unified percept. The hypothesis that conscious processing requires integrating information across multiple levels of the sensory hierarchy is by no means unique to Lamme.[3] It is a near consensus among consciousness researchers that conscious processing is marked out by the integration of information through long-range, dynamic patterns of feedback across more or less widely distributed areas of the brain. In predictive processing, this integration of information is understood as an inferential process. The brain combines information across the sensory hierarchy in such a way as to infer a coherent hypothesis that does the best job of explaining away current prediction errors.

Recall that the generative model is made up multiple interacting layers with lower layers modelling fast changing events and features of the environment such as motion, or edges at an orientation. Higher layers of the hierarchy, by contrast, model more persisting and invariant features of the environment. The layers of the hierarchical generative model are mutually constraining. The predictions that are made at higher levels help to resolve ambiguities at the lower layers. For instance, suppose the posterior at a higher-level identifies an object as an apple. This will then make highly probable certain predictions at the lower levels concerning the sensory features of the object, such as it size, shape, colour, and smell. If a prediction error arises at a lower layer in the hierarchy – perhaps the apple has a sheen like a painted piece of artificial fruit – this will then lead to an updating of the high-level hypothesis that the object is an apple.

A coherent percept is thus the result of combining top-down prior expectations formed on the basis of statistical learning with bottom-up sensory information. The brain infers the hypothesis that makes the best overall sense of current sensory information and that thus does the best job of explaining away current prediction errors. Conscious perception is the "upshot" of unconscious perceptual inference (Hohwy 2013, p. 201). The winning hypothesis – the one that gets to make it into a person's consciousness – is the one that does the best overall job of bringing everything together into a globally coherent experience.[4]

The winning hypothesis is described by Hohwy as the posterior hypothesis that best explains current sensory evidence (Hohwy 2013). The winning hypothesis emerges by finding a way of updating priors across the sensory hierarchy as a whole based on current evidence that does the best job of quashing prediction errors. We doubt, however, that casting predictive processing in terms of inference to the best explanation is the best way of making sense of what brains do. We agree with Clark that action is for the control of perception (Clark 2016, 2017b). Thus, there is no perception without action. Perceptual inference is the process of updating predictions about how to minimise surprise through action. What the organism is aiming to do in minimising prediction error is maintain itself in the sensory states it expects to be in – for instance, the sensory states it has come to associate with reward. The world that shows up for us in perception is for the most part a world of organism-relevant possibilities. The organism acts to bring about the outcomes it prefers. Priors are better thought of as preferences (Friston et al. 2015a; Friston et al. 2013). The agent acts with the aim of minimising the divergence between the outcomes that are likely given its current states and the outcomes it prefers because they are in some way valuable to the organism. In the final section of this chapter, we use this account of priors to push back against the internalist and individualist account of cognitive assembly at it applies to conscious experience.

5.5 The argument for extended consciousness from distributed cognitive assembly

Priors do not aim at mirroring a mind-independent, agent-neutral world (Clark 2017b). We have been arguing that priors are better thought of as the agent's abilities for effectively engaging with the environment. Consider again our earlier example of recognising our school friend whom we have not seen for many years. We do not simply perceive a familiar face; we are also ready to engage our long-lost friend in conversation, to catch up on what has happened since we last met. As Williams nicely puts it quoting Rorty (1979), the generative model is "for coping, not copying" (Williams 2017, p. 6). It does not represent invariant properties of an objective ready-made world. The regularities and invariants that the generative model tracks are the regular and persisting patterns of practical engagement with the environment. The stable and persisting regularities that our brains are able to track are made through our activities, in particular through the activities people engage in when they take part in social and cultural practices. They are not patterns and regularities that preexist our activities and that our brains passively mirror by maximising the evidence for the internal models they build.

We are thus much in agreement with Gallagher and Allen's proposal to conceive of predictive processing as "predictive engagement" (Gallagher and Allen 2016). Prediction error minimisation as it unfolds in the brain is part of a larger process of the dynamic adjustment of the organism to its environment. What the organism is aiming for in keeping prediction error to a minimum is the maintenance of "ongoing attunement with the environment" (Gallagher and Allen 2016, p. 8, c.f. Bruineberg et al. 2016; Kirchhoff 2017). Neural networks come to exhibit regular patterns of neural firing because of the activities people repeatedly engage in. These regular patterns of neuronal firing serve to attune the organism to regularities and possibilities of its environment.

Once we think of predictive processing as predictive engagement this opens the door to a different view of the emergence of a coherent and meaningful percept. We have been describing the precision-modulated process of integration of information across the sensory hierarchy as a process that unfolds entirely within the individual. However, if predictive processing is a matter of maintaining the organism's attunement with the environment, this restricted focus on the individual is telling us about only a part of a larger dynamic process. Priors do not penetrate perception from the inside of the individual. It is our ability to maintain attunement with the regularities in a cultural practice that can (depending on the context) exert a powerful influence on the perceptual processing of a sensory signal.

Consider in this light the well-known picture of the camouflaged Dalmatian (Figure 5.2). When you first look at this image, you see only lots of black and white blobs. But once you know what to look for – once you know how to direct your gaze – the Dalmatian immediately pops out. The seemingly random arrangement of black and white blobs takes on a meaning. We could describe this transition that takes place in experience in terms of the agent acquiring a belief – the belief that the picture depicts a Dalmatian. This belief then penetrates perception. In our preferred gloss, however, the person draws on their skill for engaging with pictures. Pictures are peculiar kinds of objects; they are objects made to depict something. In engaging with them, we use them to stand in for whatever they depict (Noë 2012, ch. 5). We are able to look "through" pictures to see whatever they depict. They communicate something to us, but to pick up on what this is requires special skills of looking and distributing attention, among other skills. The change that takes place in one's experience of the blobby picture is the result of one's learning how to engage with this picture. One acquires the ability to direct one's gaze in the right sort of way to gain access to what is depicted. At first, there is a lack of attunement with the picture. One explores further until one eventually acquires the abilities to attune to what it depicts.

Figure 5.2 Hidden image of a Dalmatian – first you see only black and white blobs, but once you know where and how to look, the next time you encounter the picture, the Dalmatian immediately pops out

Source: Thurston, J.B. and Carraher, R.G. (1966). *Optical Illusions and the Visual Arts*. Van Nostrand Reinhold

Perception is thus not penetrated from within by top-down beliefs but is "permeated" from the outside by cultural practices (Gallagher 2018, ch. 6). As Hutchins notes,

> Cultural practices include particular ways of seeing (or hearing or feeling, or smelling, or tasting) the world. Cultural practices are not cultural models traditionally construed as disembodied mental representations of knowledge. Rather they are fully embodied skills. Cultural practices organise the action in situated action.
>
> (2011, p. 5)

Hutchins gives the wonderful example of seeing a constellation. To see a constellation requires directing one's gaze at the stars in a particular way, just like in the example of the blobby picture. We can see points of light in the sky, but to see a constellation requires seeing a whole configuration of points of light, and this takes skill. It calls for skills for attending to patterns in the night sky in such a way as to identify useful stars such as the so-called "pointer" stars in Ursa Major that allow one to find Polaris (Hutchins 2014, p. 5). It requires conceiving of points of light in the sky as stars, but this is a way of conceiving of the world that emerged only with scientific practices.

One might also make use of instruments like telescopes, but we are trained in the use of these complex instruments. It may involve discursive practices that allow one to project what Hutchins calls a "trajector" onto the arrays of points of light (2014, p. 5). Seeing a constellation turns out to be a complex skill that a person comes to possess only by knowing how to perform and take part in a variety of different practices.

One might object that this example is a special case. It clearly qualifies as an example of culturally mediated form of cognition, but it does not support the more general claim that we are needing that human action is organised by cultural practices. We of course agree that there are some features of this example that do not generalise. We nevertheless contend that the permeation of perception by cultural expectations is a more pervasive phenomenon. Consider the examples of "precision-expectation based sensory sampling" that Clark (2016, §2.7) discusses. Clark reviews research investigating patterns of gaze fixation in the performance of everyday tasks like making sandwiches or cups of tea. These studies show how people saccade in ways that predict where they expect the useful information to be found. But what counts as useful information depends on the context and the task that the agent is performing. Thus, it will prove crucial for the brain to weigh the precision of prediction error in ways that are sensitive to both context and task. For instance, in cutting a sandwich in half, people tend to look where the knife makes contact with the bread (Hayhoe et al. 2003; cited by Clark 2016, p. 67).

Now this opens the door once again to the Hutchins's thesis that cultural practices organise the action in situated action. Making sandwiches and making cups of tea are both cultural practices that depend on culture-specific preferences for eating bread and drinking tea. Clark uses this research to argue that uncertainty reduction is the "driving force behind gaze allocation and attentional shift" (2016, p. 68). We would invert this claim, however, and argue that gaze allocation and attentional shifts are constrained by cultural practices. Cultural practice is one of the driving forces behind uncertainty reduction. The constraints that come from cultural practices influence how precision is weighed in a given context and thus how uncertainty is kept to a minimum.

Consider, for instance, the finding that language directly impacts precision estimation. Language can be used to summarise the complex statistical regularities, much as we see in the brain at higher levels of cortical processing invariants patterns of firing that track long-term, stable environmental regularities. Language performs this function in an externalised written or spoken format that is apt to re-enter and modulate simultaneously the neural processing of multiple individuals. Not only does language work to summarise highly complex statistical information in a neatly packaged, compact format, but also tracking statistical relationships between words and the contexts in which they occur provides a great deal of information about semantic relationships between words, as well as relationships of meaning in domains far removed from sensory experience (Clark 2016, p. 278). Linguistic practice can provide precision expectations that come from the outside. Once they have entered the system, they help to boost a prediction error signal that might otherwise have been too noisy to notice.

We see an example of this in continuous flash suppression. In this experimental paradigm, one eye is shown a static image of a face, while the other eye is presented with rapidly changing series of images (e.g., geometrical figures). The static image is suppressed from consciousness (i.e., it is not consciously detected). Hearing a word can suffice to unmask an otherwise invisible image when the image matches the named category (Lupyan and Ward 2013). Hearing the word "zebra," for instance, would enable subjects to see an image of a zebra that they would have otherwise been blind to. What hearing the word "zebra" seems to be doing in this case is creating a prior expectation, a prediction for zebras. This prior expectation then attunes the subject to zebra images.

Consider as a further example the study by Thierry et al. (2009), in which EEG was used to measure the visual mismatch negativity response (vMMN) when speakers of Greek and English were given the task of spotting an oddball stimulus. The subjects were presented with a sequence of squares with one circle introduced in between or a series of dark blue circles interspersed with one light blue circle. In the speakers of Greek, a vMMN was found for the light blue contrast that was not found in the speakers of English. The researchers hypothesise that this is because Greek contains different words for light blue (*ghalazio*) and dark blue (*ble*). Exogeneous attention (investigated using vMMN task) is differentially sensitive in speakers of Greek and English. Bottom-up prediction errors are assigned high precision and thus get to influence processing based in part on the language one speaks and the distinctions (in this case between colours) that are made in one's language.

Maintaining attunement with the cultural world depends on context-sensitive precision weighting. We can see this by considering pathological cases of precision estimation such as are hypothesised to occur in schizophrenia and autism spectrum disorder. People with schizophrenic delusion have been hypothesised to have a high expectation of noise and uncertainty. They expect more sensory noise than there really is in a given context, with the consequence that they are unable to find the signal among the noise. This leads them to dampen prediction error and ignore evidence in favour of their prior expectations (Fletcher and Frith 2009; Hohwy 2015). The effect of this aberrant weighting of precision and general failure of context-sensitive updating of precision expectations is that they come to inhabit a delusional reality that is increasingly cut-off and removed from the common-sense everyday familiar reality they share with other people (Sass 1994). They increasingly come to inhabit their own solipsistic reality.

People with autism also update precision expectations in ways that can prove pathological. Autism spectrum disorder is characterised by a wide range of sensorimotor deficits combined with social-cognitive difficulties in tracking the mental states of other people. Recently, it has been hypothesised that their sensorimotor deficits may be explained by expectations of overly precise prediction error. This leads the sensory signal reporting prediction error to dominate in processing, so people with autism then have difficulties in attuning to more stable and persisting regularities (Pellicano and Burr 2012; Lawson et al. 2014; Palmer et al. 2017).

There is clearly more to be said about psychopathology and the differences in conscious experience of the world it entails. We mention these examples, however, to highlight the importance of precision expectations for the experience of the everyday, familiar world that people ordinarily take for granted. Maintaining attunement with the world depends on setting precision expectations in ways that align with wider practices. So long as the individual fails to do so, the result is a more global lack of attunement to the everyday world in just the way we find in the examples of psychopathology we've just described.

Roepstorff and colleagues note in passing that it is not only prediction error that the enculturated brain uses as feedback for tuning the generative model (Roepstorff et al. 2010). Also important is *match* in expectations *between* people. If we are to understand the coordination that we find among people that share and take part in the same practices and that come to construct common worlds, they write, then

> [I]t may not be enough to understand how mismatches are generated and propagated (i.e. prediction errors). It appears equally important to understand how matches – with classifications and expectations – may propagate and resonate both inside brains (Grossberg 2007) and in the interaction between people (Roepstorff 2008; Tylen et al. 2009) embedded in a material world (Clark 2006).
>
> (2010, p. 1057)

Such a match in expectations and classifications matters because the result of this match is the construction of a shared cultural world. We have been arguing that particularly crucial is that people match in their precision expectations. It is only if people match in their precision expectations that sensory signals will be integrated with top-down expectations across the sensory hierarchy in such a way as to lead to shared experience across subjects. Subjects that set precision according to their own individual standards fail to gain access to such shared modes of experience.[5] By having precision expectations set in particular contexts by patterns of cultural practice, the result will be that individuals come to share experiences and ways of making sense of the world. They construct a shared lived reality through their patterns of cultural activity.

5.6 Summary

The chapter has taken up two challenges from predictive processing that seem to challenge an inference from sensorimotor enactivism to extended consciousness. The first objection that we considered showed how to understand sensorimotor understanding in predictive processing terms. We showed how the distinction between the generative model and the generative process could block such an argument. The generative process needs to be understood in wide, environment-involving terms, just as the sensorimotor enactivist supposes.

The second objection that we have taken up takes the predictive processing account of cognitive assembly in terms of precision expectations to imply internalism about the processes that lead to the construction of conscious experiences. The aim of our argument has been to show that cultural practices can do at least some of the work of dynamically configuring the cognitive processes that result in our conscious experiences. Both the constraints of cultural practices and the coalitions of neurons that softly assemble in the individual person's brain are elements of a single adaptive dynamical system.

We have shown how there are high-level expectations set up in patterned practices that come to guide cycles of perception and action (Roepstorff et al. 2010, p. 1056). These high-level expectations orchestrate the flow of information within the brain, settling the relative influence of top-down expectations and incoming sensory signals. They are high-level expectations that are set, however, by stable and regular ways of going on in practice that people partake in.

The result of setting precision based on cultural practice is the match in predictions across subjects. Through matching predictions, people come to share realities physically, mentally, and socially in interaction (Roepstorff et al. 2010, pp. 1056–1057). The relevant system for bringing about such a match in expectations is not the individual but the plurality of individuals engaged in interaction that together enact a patterned practice. We can think of the models of interacting individuals as a shared model when the models are reciprocally aligned.

We have thus shown how cultural practices may play a role in organising and assembling the neural processes that construct online conscious experience. When cast in terms of the Markov blanket formalism, these arguments suggest that it is possible to extend the Markov blanket that bounds the states that generate conscious experience to include the patterned regularities of cultural practices (c.f. Kirmayer 2018).

Notes

1 Going in a similar direction to this example, there is now substantial evidence for cross-modal influences on what a person experiences where experience in one sense modality (e.g., vision) influences what one experiences in a different sense modality (e.g., taste). Thus, seeing a white wine that has been coloured red suffices to change subject's gustatory experience of the wine. Subjects describe the wine as tasting more like a bottle of red than white (Shankar et al. 2010, discussed by Clark 2016, p. 55).

2 We use the notions of "belief," "expectation," and "knowledge" interchangeably in this section to refer to the prior "beliefs" that are embodied in the generative model. For a more careful discussion, see Chapter 3. Exactly how this notion of belief and knowledge maps onto the folk-psychological usage of these terms is controversial. We are in no way committed to a one-to-one mapping of personal-level doxastic or belief-like states onto the dynamics of the generative model. For some discussion of this issue, see Macpherson (2017, p. 12) and Drayson (2017, p. 8).

3 See, e.g., Dehaene and Changeux (2011); Melloni and Singer (2010); Edelman and Tononi (2001); Thompson and Varela (2001). Lamme's novel twist on this hypothesis is his association of conscious processing with perceptual organisation, which he takes to be implemented in the brain by recurrent cycles of neural processing in sensory cortex. Other theories take large-scale synchronisation to be the key neural mechanism that delivers integration of information across the brain. We argue for the importance of multiscale integration in Section 6.2.2., when we develop a diachronic constitutive account of predictive processing and its relation to consciousness.

4 Hohwy has used this account of consciousness to make sense of why consciousness might be associated with what Dehaene and Changeux (2011) describe as the "ignition" of large-scale prefrontal-parietal network. Global ignition takes place when a hypothesis emerges that is able to explain away current sensory evidence and thus has the highest posterior probability of the available hypotheses. Such a hypothesis is then broadcast to other systems across the brain to play a role in the guidance of planning and in decision-making and to inform subjective report.

5 This results in the loss of basic trust in the everyday world that is a signature feature of the prodromal stage in the onset of schizophrenia and that may be the source of everyday anxiety that is a pervasive feature of everyday experience in people with autism. For further discussion of this loss of basic trust in terms that make a good fit with predictive processing, see Ratcliffe (2017).

6 Extended diachronic constitution, predictive processing, and conscious experience

6.1 Introduction

In this final chapter, we take up a challenge to the argument from distributed cognitive assembly of the previous chapter. There we argued that the context-sensitive assembly of the neural processes involved in minimising expected long-run prediction error is organised in part by factors outside of the individual agent. The precision estimating dynamics that determine the influence of prediction error signals can be tuned and constrained by the regularities enacted in cultural practices. We argued that cultural practices not only organise the action in situated action but also play a constitutive role in the self-assembly of processes that work to minimise prediction error in the here and now. Thus, the role of the precision mechanism in the context-sensitive assembly of neural processes that lead to consciousness does not provide support for internalism. It is better interpreted as lending support to a third-wave extended mind perspective on the realisers of conscious experience. The weighing of the precision of a prediction error signal in a given context is mediated by an extended network of causal factors that loop out into the cultural environment.

Someone sceptical of this argument might still ask, is consciousness not constituted by *internal* brain states in the here and now? This sceptic may agree with us that patterns of cultural practices are involved in setting the parameters of internal generative models over long timescales. They may nevertheless insist that in the here-and-now conscious experience is fully determined by a neurally instantiated generative model. In Clark's (2011) reply to Hutchins's (2011) hypothesis of enculturated cognition, he presented an argument along these lines. He argued that cultural practices operating over long-term timescales *merely set the scene* for "variously neurally-based processes of cognitive assembly" (2011, p. 458). So when it comes to explaining what assembles cognitive processes to deal with on-the-spot challenges, Clark's view is that the brain accomplishes this task.

Clark's (2008) hypothesis of organism-centred cognition can thus be marshalled to mount an objection against our argument from distributed cognitive assembly. The worry is a version of the causal-coupling-constitution fallacy for the extended mind. Cultural practices, it is objected, can at best causally influence the processes of developing the neural architecture for dealing with on-the-spot prediction error minimisation. The regular patterns of activity found in cultural practices cannot make a constitutive contribution to prediction error minimisation at a given instant in time. So, patterned practices cannot be a part of the machinery that accounts for the constitution of conscious experience.

Our aim in this final chapter will be to develop a response to the causal-coupling-constitution objection as it applies to the third wave. We will do this by explaining why, in our view, the constitution of consciousness by predictive processing must be understood in diachronic, and not merely in synchronic, terms. We will argue that the constitutive basis of online perceptual experience involves attunement between internal and external dynamics. This ongoing attunement is realised in a network of processes that comprises both a past history of engagement with the world and a future-oriented attunement to a constantly changing environment. Such a diachronic view of constitution implies that perceptual experience cannot always be constituted by internal neural processes.

The structure of this final chapter is as follows. We start, in Section 6.1, providing a general overview of the concept of diachronic constitution. To make sense of the constitution of consciousness by predictive processing, we to pay special attention to temporality because consciousness is inherently backwards and forwards flowing in its temporal structure. In Section 6.2, we return to the causal/coupling-constitution fallacy as it applies to the third wave. The notion of diachronic constitution provides us with the conceptual tools needed to decisively meet this challenge. We show that once one thinks of predictive processing as constituting conscious experience diachronically, one is required to sometimes look beyond the brain for a metaphysical account of how perceptual experience is constituted. Finally, in Section 6.3, we confront some complex modal intuitions about the constitution of consciousness. We show that familiar "neural duplication" arguments for consciousness fail once one takes seriously the diachronic understanding of constitution that we develop in this chapter.

6.2 Diachronic constitution: towards a process theory of constitution

In the previous chapter, we argued that perceptual experience is partly constituted by sensorimotor expectations that guide a perceiver's skilled engagement with the world. Here we aim to show how embodied engagement with the world should be thought of in terms of a process-based ontology.

First, we must say something brief about the metaphysical notion of constitution as we use it in this chapter. The notion of constitution is standardly understood as a composition relation intended to capture part-whole relations. Shapiro (2011) illustrates what he takes "constitution" to mean by using the example of water. He says that "Oxygen is a *constituent* of water, because water consists in atoms of oxygen conjoined with atoms of hydrogen" (2011, p. 20). The relation of composition that holds between the parts and the whole is the relation of constitution. Constitution is also a metaphysical dependence relation. Long-term potentiation (LTP), for instance, constitutively depends on the gating functions carried out by the N-methyl-D-aspartate (NMDA) receptor (Craver 2007). The constitution relation here refers to the relation that holds between the parts of the systems and the behaviour of the system as a whole. Each component part carries out a function that is responsible for realising a sub-capacity. When these components are organised in the right way, the result is the constitution of the phenomenon of interest (e.g., LTP).

In this chapter, we are interested in the constitution of phenomenal consciousness by predictive processing. Thus, we are considering the metaphysical relation of constitutive dependence that holds between phenomenal consciousness – this is the phenomenon of interest – and predictive processing. Predictive processing tells us what the parts of the system must be doing such that when these parts are organised in the right way, they constitute consciousness. The parts of the system will include, for instance, components that perform predictions, error calculation, precision estimation, and so on.

Diachronic constitution is a temporalised and dynamic relation of constitution. Hence, it stands in sharp contrast to the received view of constitution in analytic metaphysics, which understands constitution as a synchronic (i.e., atemporal) relation. Constitution is typically taken to be synchronic in order to distinguish it from the relation of causation, which is paradigmatically diachronic (Bennett 2011). Our claim that constitution is diachronic blurs any strict distinction between constitution and causation (Gallagher 2018; Kirchhoff 2015b). If constitution is diachronic, it has a very different metaphysical profile from what one would expect given the received view of constitution (Table 6.1). We will unpack these differences next before returning to predictive processing.

Synchronic constitution works best in accounts of the constitution of material objects, such as tables and chairs (Kirchhoff 2015a). Unlike material objects, processes are extended in time. Processes take up time from the

Table 6.1 This table highlights the key differences between the properties of synchronic constitution (left side) and the properties of diachronic constitution (right side)

Synchronic constitution	*Diachronic constitution*
• Synchronicity	• Diachronicity
• Discrete time	• Time dependence
• Non-causal	• Circular causation
• Object based	• Process based

moment they begin to when they end. In contrast to material objects, whose identity conditions are entirely determined at each moment of their existence, processes depend on temporal unfolding to be what they are, because processes "are creatures of time" (Noë 2006, p. 28; see also Hofweber and Velleman 2011). To understand what processes are, we need an ontology that is temporal and dynamic as contrasted with the substance ontology we employ to make sense of the material constitution of material objects.

Embodied engagement with the environment is process-like in nature. As Gallagher (2017) rightly notes, embodied activity is more than the configuration of the body from moment to moment that it "involves a trajectory and is constantly on the way" (Gallagher 2017, p. 95). The precise posture of the body at a single moment is "a complete abstraction from an ongoing movement" (Gallagher 2017, p. 95). As my hand moves towards the cup of coffee I am reaching to grasp, my arm goes through a sequence of different postures. At each moment, my movement is unfolding along a trajectory because of the cup I am aiming to reach. There is thus a retaining in perceptual presence of the cup's affordances – its possibilities for action – to which my movements are coordinating and adjusting. At the same time, my movements are unfolding in a way that anticipates my taking hold of the cup of coffee to drink from it. My movements thus unfold along a particular trajectory based both on a retention of my body's configuration in relation to the environment and on an anticipation of where my movement is heading next. Similarly, perception is not a "knife-edge impression of the present." Perception arises with what Gallagher describes as an "empty anticipation" that is either fulfilled or not fulfilled. This empty anticipation is in turn constrained by what Gallagher calls "retention" of what was just anticipated.

Temporality explains the directedness of both consciousness and action towards something in the environment. Consciousness as enactive is to be understood as an "I can" that is as an "apprehension of the possibilities or affordances in the present" (Gallagher 2017, p. 97). There would be no engagement with affordances were perception to present an animal with only a knife-edge present. To apprehend and be sensitive to possibilities, a perceiving animal needs prospection: it needs to have experiences that reach out into the future, anticipating what could be. This is just what it takes to perceive possibilities. Gallagher does not spell out whether perception of possibilities would be possible without retention. However, since what is retained is just the fulfilled or unfulfilled prospection that has just past, we can infer that it would not. Perception without retention would be perception that is unconstrained by what was previously anticipated. But we have just argued that there can be no perception of possibilities without prospection.

The interesting question to ask now is, how does predictive processing capture the intrinsically diachronic nature of embodied engagement with the world?

First, predictive processing is hierarchical, involving the minimisation of prediction error over nested and multilayered temporal and spatial scales. A common observation is that microscopic dynamics predict changes in sensory states that unfold over fast timescales. Global or macroscopic processes in turn predict statistical regularities or invariants associated with states unfolding over slower timescales (Hobson and Friston 2014; Hohwy 2012). Lloyd (2018) argues that this temporal division of labour is perfectly suited for explaining how it is possible, at any given moment in time, to experience the world as a set of not-yet-actualised possibilities (c.f. Gallagher 2018; Madary 2016). We think this is correct because predictive processing provides a how-explanation of this enactive feature of conscious experience in terms of diachronic or temporally deep generative models – the kind of models we considered in the previous chapter as supporting counterfactual prediction. In discussing the temporally deep generative model, Friston says that

> systems that can grasp the impact of their future actions must necessarily have a *temporal thickness*. They must have internal models of themselves and the world that allow them to make predictions about things that have not and might not actually happen. Such models can be thicker and thinner, deeper or shallower, depending on how far forward they *predict*, as well as how far back they *postdict*, that is, whether they can capture how things might have ended up if they had acted differently. Systems with deeper temporal structures will be better at inferring the counterfactual consequences of their actions.
>
> (2017, *Aeon*)

"Temporal thickness" refers to the capacity to make inferences not only about the present but also about the past and the future. A temporally deep model will allow an organism to project itself through time. Such an organism will be able to select actions that minimise its expected or future uncertainty. It will able to engage in "proactive, purposeful inference about its own future" (Friston 2017, *Aeon*).

Why is it that the predictive architecture that is found in conscious creatures has a temporally deep hierarchical organisation? The answer to this question lies in the *time-dependent* and *circular causal* dynamics involved in predictive processing. Hobson and Friston (2014) have proposed a predictive processing account of consciousness premised on these properties of circular causality and time dependence. They claim that "low-level inference of the sort associated with motor reflexes. . . does not itself constitute conscious processing until contextualised by deep hierarchical inference at higher levels [in the predictive hierarchy]" (2014, p. 6).

This is significant for a couple of reasons. The first is that it underpins the enactive view that were an organism to be unable to engage with the experienced world as a set of possibilities, it would not be an organism capable of being conscious.[1] There is therefore a deeply probabilistic aspect to consciousness. The second is that deep temporal models have temporal thickness because the dynamics involved in prediction error minimisation span different temporal frequencies. As we mentioned above, low-level predictions target changes in sensory states over fast timescales, while high-level predictions target statistical regularities or invariants associated with states unfolding over slower timescales. These differences in temporal scale correspond to differences between microscopic (local) and macroscopic (global) prediction error minimisation.

The final thing we want to address is the *circular causal interactions* between local and global prediction error minimisation. In predictive processing, circular causality captures the idea that macroscopic dynamics are causally relevant for prediction error minimisation at lower scales in the predictive hierarchy. There is nothing unusual about this proposal. The idea that global dynamics constrain local dynamics is a common feature of self-organising dynamical systems (Hobson and Friston 2014). Indeed, Hobson and Friston refer to this particular feature when they say that high-level predictions contextualise low-level predictions such that this top-down effect from global dyanmics to local dynamics in part determines conscious experience.

Gallagher has recently argued that the temporal depth of embodied engagement with the world should be analysed by using a diachronic notion of constitution. He says that "cognitive processes occur on several timescales" and that it is "important to grasp how these timescales interrelate in order to understand how constitution can be, not only diachronic, but also non-linearly dynamic" (2018, p. 9). We agree with Gallagher on these points (Kirchhoff 2015a, 2015c). Related ideas are foreshadowed in our earlier observation that prediction error minimisation is not only temporal but also composed of multiply nested and varied processes that work together to minimise long-term expected prediction error. Our proposal to understand the constitution of consciousness in terms of predictive processing therefore has much in common with other proposals about diachronic constitution in the enactivist and phenomenological literature.[2]

It should now be clear why we think of the account of how consciousness is constituted by predictive processing as calling for a diachronic account of constitution. In the next section, we address the causal-coupling-constitution fallacy as it arises for third-wave extended mind. We do so first by restating Clark's challenge to our argument from distributed cognitive assembly. Our strategy will be to argue that given its inherently temporal processing profile, one should acknowledge that prediction error minimisation is diachronically constituted. It is then a small step to also acknowledge that the self-assembly of processes needed to minimise prediction error is also on occasion extended.

6.3 The causal-coupling-constitution fallacy in third-wave extended mind

Clark's (2008) picture of cognitive assembly is firmly internalist: a causal and self-organising arrow goes from the organism to elements in the environment. It is the canny individual in his view that establishes and maintains the strings in a larger net of prediction error minimisation. Over the course of this book, we have argued that while this organism-centred account might sometimes be appropriate, it fails to do justice to the dynamic role played by cultural practices in the overall project of prediction error minimisation.

There are several reasons why we think Clark ought to update his account of cognitive assembly. Indeed, there are also signs in his writings that he recognises the need for such an update. Consider, for instance, the following somewhat surprising quote in his response to Hutchins's critique of Clark's (2008) book: "There is, in short, a kind of 'family business' here in which the brain, the body, and the self-structured (at multiple timescales) environment all pitch in" (2011, p. 454). In the preceding sentences, Clark even embraces the idea that the "overarching goal of minimizing informational surprise can also be served. . . by the canny longer-term structuring of an environment as when we write down our ideas while thinking, put signs on shops, paint arrows on country walks, and so on" (2011, p. 454; c.f. Clark 2013, 2016, ch. 9). If all these different informational dynamics can be part of an extended network of processes that minimise long-term prediction error, why then restrict the self-assembly of such networks to the brain?

One reason might turn on the meaning of "pitching in." Some of the slower, long timescale processes might do the work of getting the assemblies of prediction error minimising networks off the ground and up and running without themselves doing the work of assembling the system. We suggest that this is most likely Clark's position, even if he is willing to entertain the alternative family business model. However, we think that there are good reasons not to succumb to this retrograde throwback to internalism.

Clark's organism-centred view of cognitive assembly takes the causal flow of influence to travel in one direction: from the individual out into the environment. This misses the way in which the flow of causal influence can be bidirectional because of the dynamical nature of the cultural environment in which people act. The day-to-day routines of individuals include cultural practices such as dancing, talking, writing, voting, queuing, listening to music, driving, and so on. These are activities that unfold when individuals coordinate with one another and do things together.

Consider, as a stripped-down example, studies that show how infants configure and adjust their behaviour, including emotional behaviour, conditioned on the social and emotional responsiveness of their caregivers. It is now a well-established result that disrupting the synchronised bodily expressions between infant and caregiver, such as occurs when a caregiver adopts a "still-face," results in signs of distress on the part of the infant (Adamson and Frick 2003). The infant expects the caregiver to play their part in the unfolding proto-communicative exchange. The caregiver is expected to take their turn. The failure to play along and do what the infant expects is the source of distress. When the caregiver reinitiates her responsiveness, the "effects of the [interaction] disturbance persist as a spillover" (Nagy 2008, p. 1782). This kind of disruptive intervention in a mutual interaction leaves the infant with an experience of distress. The failure of the caregiver to play their part in the interaction leads to misalignment and a breakdown in communication, perhaps even to a temporary loss of trust. What this example highlights is that other people often play an active and dynamic part in the looping cycles of perceptual engagement with the environment. In the interactions that unfold between people, shared expectations take shape that the individuals taking part in the interaction enact. As Friston and Frith note,

> if there is a shared narrative or dynamic that both brains subscribe to, they can predict each other exactly, at least for short periods of time. . . when two or more (formally similar) active inference schemes are coupled to each other. . . the result of this coupling is called generalized synchronization.
>
> (2015, p. 2)

Clark will no doubt continue to insist that even if social interaction and cultural practice is a part of the ultimate (evolutionary, developmental) cause of prediction error minimisation, the "proximal cause (the mechanism) of larger amounts of surprisal reduction" is "the operation of a cortical processing regime" (2013, p. 235). One should not mistake an ultimate for a proximal explanation. Indeed, the causal-coupling-constitution fallacy in third-wave extended mind means ignoring this distinction and making the mistake of treating ultimate causes that lie in the agent's past as if they were a part of the proximate mechanism that account for behaviour in the here and now.

We think this line of argumentation misses the extent to which prediction error minimisation depends on maintaining ongoing attunement to the environment. Consider the phenomenon of culture shock as a case in which such attunement fails. In culture shock, subjects describe feelings of distress and alienation, probably due to high and persistent prediction error. Their prior expectations about sensory observations fail to attune with the actual observations conditioned on the cultural practices in their new home. A much-discussed case is that of 13-year old Eva Hoffman, who, along with her mother and father, left Poland in 1956 for the prospects of a better life in Vancouver, Canada. Even though Eva had her parents by her side, her experiential world changed dramatically. She explains:

> the country of my childhood lives within me with a primacy that is a form of love. . . It has fed me language, perceptions, sounds. . . It has given me the colors and the furrows of reality, my first loves.
>
> (Hoffman 1989, pp. 74–75; quoted in Wexler 2008, p. 175)

Having spent only three nights in Vancouver, she reports waking up from a dream, wondering

> what has happened to me in this new world? I don't know. I don't see what I've seen, don't comprehend what's in front of me. I'm not filled with language anymore, and I have only a memory of fullness to anguish me with the knowledge that, in this dark and empty state, I don't really exist.
>
> (Hoffman 1989, p. 180; quoted in Wexler 2008, p. 175)

Wexler explains in *Brain and Culture* this experience of alienation: as "she attempts to take in her new environment the requisite internal structures are lacking or the old structures are obstructing" (2008, p. 176).

Patterned practices can thus be a part of the constitutive basis of conscious experience in the proximate here and now, as is illustrated by the example of culture shock. They guide and constrain what the person experiences and how they act in the here and now. We have been arguing that the higher levels of the individual's generative model serve to attune them to the regularities of the cultural environment. Expectations that have their origin outside of the individual in patterns of practice come to be thoroughly integrated into the generative model. Should the individual move to a new environment, the result will be misalignment and pervasive prediction error that is hard to suppress. The priors that are operative in the practices found in this environment fail to match the priors that were operative in their old environment. The result is that their perception and action are guided by expectations that no longer fit with those of other people. In the here and now, it is the misalignment between priors and contextual regularities that gives rise to culture shock. This kind of misalignment arises through the present relationship between a person's internal dynamics (including the social expectations that are embodied at the higher levels of their model) and the regularities of their new niche.

The case of culture shock thus suggests that prior expectations are acquired and updated by ongoing participation in a cultural niche. Once we take seriously Eva's current (Canadian) niche, can we continue to maintain that it is the neural machinery that fully constitutes and explains why she has the kind of experiences she does? Just as it makes no sense to look for what Hurley calls an "internal end-point" after which we can throw away the world, so it is with practices – there is no external endpoint. There is no point at which the expectations that form in a practice become fixed once and for all. What is expected of participants is continuously being settled and is a matter of ongoing negotiation by participants in practices.

As we argued earlier, the internal and the external stand in a relation of reciprocal dependence. They are not separated and secluded from each other by the boundary that gets established by the Markov blankets. Moreover, just as the internal dynamics are continuously undergoing change based on an imperative to keep prediction error to a minimum, so too are external dynamics. The external dynamics found in the environment (which are shaped by the things that people do when acting together in practices) are continuously undergoing change to fit and accommodate what individuals do.

Thus, to explain culture shock, we need to appeal to an extended dynamic singularity comprising Eva's internal states, the patterns of practice that are enacted within her cultural niche, and the generative process coupling her to the environment. To explain her current experiences, one cannot help taking into consideration her involvement in past cultural practices and their role in maintaining the priors comprising her generative model. This is because priors are tuned and updated by prior cultural learning. But emphasising this extensive prior learning history at the expense of here-and-now practices is a mistake. It is not just her past that we need to take into account but also her present circumstances and her orientation to the future in her new cultural environment.

Her experience is diachronically constituted. The constitutive basis of her experience in the here and now involves the regularities in cultural practices. It constitutively depends on *both* the priors established in her past (in Poland) and those exhibited in her present and ongoing situation (in Vancouver). Thus, one cannot simply screen off as background conditions the patterned interactions between individuals in cultural practices when addressing the constituents of prediction error minimisation.

Our response to the causal-coupling-constitution fallacy is thus that the individual agent is a locus of coordination of neural, bodily, and environmental elements. The coordination of these elements is sometimes organised by the normative constraints that operate in the cultural practice. We saw in Section 6.1 how backwards or top-down connections can constrain and modulate forward connections or bottom-up connections. The dynamics at lower hierarchical scales of operation predicts statistical regularities whose dynamics have high variability and unfold over fast timescales. At the higher layers of the processing hierarchy, we find macroscopic dynamics attuned to regularities whose dynamical properties unfold over slower timescales. We showed how the dynamics at these macroscopic scales modulates the dynamics at microscopic scales in a global-to-local fashion. The higher layers of the processing hierarchy attune an individual to regularities in the cultural practice, making it likely that people align with each other in their understanding of the world. As was argued in the previous chapter, people are able to enjoy shared experiences of the world only because their predictions are conditioned on matching high-level expectations. We suggest, then, that what we find at the top of a hierarchically structured generative model are shared expectations that attune individuals to patterns of cultural practice. All of this suggests to us that the patterns of expectancy that take shape in cultural practices play a constitutive role in prediction error minimisation. The notion of constitution needed for understanding how predictive processing constitutes perceptual experience of the world is both diachronic and extended.

6.4 Modal intuitions: the neural duplicate intuition and why it is wrong

We will end this chapter by considering an objection also discussed at length in Hurley (1998) and Noë (2006). Would my neural duplicate here and now not have the same experience as I have had? Alternatively, would the neural duplicate of Eva (from the culture shock case above) in the present moment not have the same phenomenal experience as Eva just now?

One way to address this question is to ask, under what conditions could Eva and Eva* be neural duplicates? Or, under what conditions could Eva and Eva* be duplicates in the sense of instantiating the same generative model? Under predictive processing, we come to be the generative model we are in virtue of how we influence the external environment and how the environment influences us. So for Eva and Eva* to actually be neural duplicates, they must also be environmental duplicates. Furthermore, they must be duplicates of their overall phenotypic states and traits – that is, they must share the same distribution of functional and physiological states, morphological features, patterns of behavioural dispositions, and ecological, social, cultural, and material niche that makes them the kind of individuals that they are. Following Noë (2006), let us say that Eva's neural duplicate Eva* has the same phenomenal experience as Eva at time *t*. We would add (in line with Noë) the observation that imagining this type of duplication seems (nomologically) possible only because Eva's neural duplicate is "embedded in and interacting with a duplicate of [Eva's] environment" (2006, p. 18). This suggests that the mere fact that Eva and Eva* are neural duplicates is not sufficient for them to be phenomenal duplicates – they must be phenotypical duplicates in the extended sense of a phenotype.

The ever-persistent internalist will no doubt object: surely Eva* can have the same phenomenal experience as Eva, whatever differences there might be between their respective environments. The internalist intuition here is familiar: once we fix the neural contribution to consciousness, variation in the environment of the individual is beside the point. The phenomenal experience of Eva and Eva* is fully settled by whatever is taking place within their respective brains. Eva* could just as well be a disembodied brain floating about in space, because all that matters when it comes to her phenomenal experience is the configuration of neural activity in her brain.

Like Noë, we shall not pretend to know the final answer to this question. There may be, at the outer limits of this modal intuition, a possible world where Eva and Eva* could share the same phenomenal experience despite living in different environments. But we can limit the scope substantially to just the present example of culture shock. We need only a single case of extended consciousness to combat the intuition that is being pumped in this thought experiment.

In the culture shock example, we argued that ongoing interactions in cultural practices cannot be ignored or sidelined as mere causal background conditions. We further argued that generative models have the attributes that they do in virtue of an agent's history of engagement with their local environment. We took this to justify the claim that perceptual experience is diachronically constituted in an extended dynamic singularity. These two lines of argumentation suggest to us that Eva and Eva* could not share the same experience of deep alienation simply by fixing their generative model and allowing for variation in their environment.

To push this further, imagine a scenario in which Eva and Eva* are in identical neural states. They instantiate the same generative model and are generating exactly the same predictions. Eva, however, is living in her home country, Poland, whereas Eva* has left to take up a new life in Vancouver. In this example, it is Eva* and not Eva that has the experience of alienation. What can we appeal to for an explanation of Eva*'s experience of culture shock? The modal intuition above suggests that Eva and Eva* can have the same experience of culture shock despite being in different social and cultural environment. This is a logical possibility but not a very plausible one. Nor is this modal possibility informative about matters in the world that we care about when providing naturalistic and philosophical explanations in the sciences of mind. The key difference maker is cultural practices. Cultural practices are being intervened on in this imaginary example, and they make the real explanatory difference in accounting for why Eva and Eva* could be generative duplicates and still differ in their phenomenal experience at a specific moment in time. Eva* experiences culture shock, not Eva. The difference maker is the change in the local environment conditioned on a developmental history of engaging in a particular suite of cultural practices. There is thus (c.f. Hurley 2010) little reason (besides internalist intuitions) to think that you can hold the internal states of the agent constant while varying the external states of the environments without this affecting what the subject experiences. As Hurley puts it, "In some cases, internal factors may not be unpluggable and repluggable across near worlds so that internal and external factors are not explanatorily separable" (2010, p. 122).

6.5 Summary

Over the course of our book, we have advanced the claim that it is the embodied agent that comprises the generative model. Organisms tend to frequent a bounded or limited set of states. These states are in part defined by the organism's phenotype. It is the repertoire of various states, ranging from functional and adaptive states to states of the environment, that makes an organism the kind of organism that it is (Friston 2011; Bruineberg and Rietveld 2014; Kirchhoff et al. 2018; Ramstead et al. 2017; Kiverstein 2018). This means that we are the kind of creature that we are in large part due to the cultural niche that we inhabit, and how we adapt to the dynamics of the niche over different timescales through the generative process (Friston 2011). The form of our generative model – that is, its parameters and action policies – come to reflect the causal, temporal, and statistical structure of the embedding environment, which in the case of humans is cultural in nature. People and the cultural niches they build around them are co-constructed. People exist in culture. Culture is not something outside and external to us.

The appeal to diachronic constitution helps us to go beyond attempts to give a special role to either the brain, the body, or the environment. The notion of diachronic constitution paves the way for thinking about the constitution of conscious experience as multilayered and nested across different scales of systemic organisation. We have been arguing that the external orbits of an extended and diachronic dynamic can be partly constitutive of conscious experience.

Notes

1 For further discussion, see Kiverstein (2018).
2 We go beyond Gallagher's work on diachronic constitution in several respects. First, Gallagher does not consider diachronic constitution in relation to prediction processing as we do. Second, even though he gestures at the idea of *extended* diachronic constitution, this idea is not developed in Gallagher's work.

Concluding remarks

We have argued that the boundaries of the conscious mind are flexible and negotiable. Our phenomenally conscious experiences are not confined to the inside of our heads but are instead constituted by activities of directly engaging with and exploring the world, at least for much of our waking experience. Phenomenal consciousness opens us to the world. We will conclude our book by briefly reviewing how the three theoretical strands to our argument have been woven together. The three theoretical strands to repeat are sensorimotor enactivism, predictive processing, and third-wave extended mind.

Predictive processing is best understood in the terms of third-wave extended mind. Predictive processing identifies the mechanisms and processes that explain how phenomenal consciousness is constituted. It follows that the mechanisms and processes that constitute phenomenal consciousness are both temporally and spatially extended, as is claimed by the DEUTS argument. Predictive processing has thus been used in our book to unpack the mechanisms and processes that the DEUTS argument claims deliver extended phenomenal consciousness.

We have argued for this conclusion by defending the following four theses:

1. The ongoing tuning and maintenance of the generative model by the generative process entails an extended dynamic singularity. It entails the dynamic entanglement of the agent and environment.
2. There is no single, fixed, and permanent boundary separating the inner conscious mind from the outside world. The boundary separating conscious beings from the outside world is fluid and actively constructed through the process of forming and maintaining the nested structure of Markov blankets that separates what is inside the agent from what is outside.

3 The predictive processes that lead to the construction of conscious experience do not only unfold within the individual but are mediated and permeated by cultural practice. The individual agent is thus better thought of as locus of coordination where the process of coordination is mediated by cultural practices.
4 The diachronic constitution of phenomenal consciousness through predictive processing entails the ongoing dynamic entanglement of the agent and its wider cultural environment.

Given the diachronic constitution of phenomenal consciousness by predictive processing, it follows that there is no point in the development of the processes that generate phenomenal consciousness after which this engagement with the environment is no longer necessary. It is for this reason that the processes in the brain by themselves will prove insufficient to generate phenomenal consciousness in isolation from the external environment. Phenomenal consciousness is first and foremost generated via active engagement with the wider environment.

Bibliography

Adams, F., and Aizawa, K. (2010). Defending the bounds of cognition. In R. Menary (Ed.), *The Extended Mind* (pp. 67–80). Cambridge, MA: The MIT Press.

Adams, F., and Aizawa, A. (2008). *The Bounds of Cognition.* New York: Wiley-Blackwell.

Adamson, B.L., and Frick, J.E. (2003). The still face: A history of a shared experimental paradigm. *Infancy*, 4(4), 451–473.

Aizawa, K. (2010). Consciousness: Don't give up on the brain. *Royal Institute of Philosophy Supplement* (The Metaphysics of Consciousness: Essays in Honour of Timothy Sprigge, Ed's P.-F. Basile, P. Phemister, L. McHenry, and J. Kiverstein), 263–284.

Aizawa, K., and Gillett, C. (2009a). The (multiple) realization of psychological and other properties in the sciences. *Mind & Language*, 24(2), 181–208.

Aizawa, K., and Gillett, C. (2009b). Levels, individual variation and massive multiple realization in neurobiology. In J. Bickle (Ed.), *Oxford Handbook of Philosophy and Neuroscience* (pp. 539–582). Oxford: Oxford University Press.

Allen, M., and Friston, K.J. (2016). From cognitivism to autopoiesis: Towards a computational framework for the embodied mind. *Synthese*, 1–24. doi: 10.1007/s11229-016-1288-5.

Anderson, M.L. (2017). Of Bayes and bullets: An embodied, situated, targeting-based account of predictive processing. In T. Metzinger and W. Wiese (Eds), *Philosophy and Predictive Processing* (Vol. 3). Frankfurt am Main: MIND Group. doi: 10.15502/9783958573055.

Anderson, M.L. (2014). *After Phrenology: Neural Reuse and the Interactive Brain.* Cambridge, MA; London, England: The MIT Press.

Anderson, M.L., and Chemero, T. (2013). The problem with brain GUTs: Conflation of different senses of 'prediction' threatens metaphysical disaster. *Behavioral and Brain Sciences*, 36, 204–205.

Anderson, M.L., Richardson, M.J., and Chemero, A. (2012). Eroding the boundaries of cognition: Implications of embodiment. *Topics in Cognitive Science*, 4(4), 717–730.

Ashby, R.W. (1960). *Design for a Brain.* New York: John Wiley & Sons.

Bach-y-Rita, P., and Kercel, W.S. (2003). Sensory substitution and the human-machine interface. *Trends in Cognitive Science*, 7(12), 541–546.

Bar, M. (2007). The proactive brain: Using analogies and associations to generate predictions. *Trends in Cognitive Sciences*, 11(7), 280–289.
Beal, M.J. (2003). *Variational Algorithms for Approximate Bayesian Inference*. PhD Thesis, University College London, UK.
Bechtel, W. (2009). Explanation: Mechanism, modularity, and situated cognition. In P. Robbins and M. Aydede (Eds), *Cambridge Handbook of Situated Cognition* (pp. 155–170). Cambridge: Cambridge University Press.
Beer, R. (2003). The dynamic of active categorical perception in an evolved model agent. *Adaptive Behavior*, 11(4), 209–243.
Beer, R. (1995). A dynamical systems perspective on agent-environment interaction. *Artificial Intelligence*, 72(1–2), 173–215.
Bennett, K. (2011). Construction area (no hard hat required). *Philosophical Studies*, 154(1), 79–104.
Bennett, K. (2004). Spatio-temporal coincidence and the grounding problem. *Philosophical Studies*, 118(3), 339–371.
Bickle, J. (2003). *Philosophy and Neuroscience: A Ruthlessly Reductive Account*. Dordrecht: Kluwer Academic Publishers.
Block, N. (2005). Review of Alva Noë's *Action in perception. Journal of Philosophy*, 102, 259–272.
Block, N. (1978). Troubles with functionalism. *Minnesota Studies in the Philosophy of Science*, 9, 261–325.
Boysen, S.T., Bernston, G., Hannan, M., and Cacioppo, J. (1996). Quantity-based inference and symbolic representation in chimpanzees (*Pan troglodytes*). *Journal of Experimental Psychology: Animal Behaviour Processes*, 22, 76–86.
Bruineberg, J., Kiverstein, J., and Rietveld, E. (2016). The anticipating brain is not a scientist: The free-energy principle from an ecological-enactive perspective. *Synthese*, 1–28. doi: 10.1007/s11229-016-1239-1.
Bruineberg, J., and Rietveld, E. (2014). Self-organisation, free energy minimisation and optimal grip on a field of affordances. *Frontiers in Human Neuroscience*. doi: 10.3389/fnhum.2014.00599.
Burge, T. (1986). Individualism and psychology. *Philosophical Review*, 95, 3–45.
Burge, T. (1979). Individualism and the mental. *Midwest Studies in Philosophy*, 4, 73–122.
Cash, M. (2013). Cognition without borders: 'Third-Wave' socially distributed cognition and relational autonomy. *Cognitive Systems Research*. http://dx.doi.org/10.1016/j.cogsys.2013.03.007.
Chalmers, D. (forthcoming). Extended cognition and extended consciousness. In M. Colombo, L. Irvine, and M. Stapleton (Eds.), *Andy Clark and His Critics*. Wiley-Blackwell.
Chalmers, D. (2008). Foreword to *Supersizing the Mind*. Oxford: Oxford University Press.
Chalmers, D. (1995). *The Conscious Mind*. Oxford: Oxford University Press.
Chemero, A. (2009). *Radical Embodied Cognitive Science*. Cambridge, MA: The MIT Press.
Chemero, A., and Silberstein, M. (2008). After the philosophy of mind: Replacing scholasticism with science. *Philosophy of Science*, 75(1), 1–27.

Churchland, P. (1992). *A Neurocomputational Perspective: The Nature of Mind and the Structure of Science*. Cambridge, MA: The MIT Press.

Clark, A. (2018). Beyond the Bayesian blur: Predictive processing and the nature of subjective experience. *Journal of Consciousness Studies*, 25(3–4), 71–87.

Clark, A. (2017a). How to knit your own Markov blanket: Resisting the second law with metamorphic minds. In T. Metzinger and W. Wiese (Eds), *Philosophy and Predictive Processing* (Vol. 3). Frankfurt am Main: MIND Group. doi: 10.15502/9783958573031.

Clark, A. (2017b). Busting-out: Predictive brains, embodied minds, and the puzzle of the evidentiary veil. *Noûs*, 51(4), 727–753.

Clark, A. (2016). *Surfing Uncertainty*. Oxford: Oxford University Press.

Clark, A. (2014). Perceiving as predicting. In D. Stokes, M. Matthen, and S. Biggs (Eds), *Perception and Its Modalities*. New York: Oxford University Press.

Clark, A. (2013). Whatever next? Predictive brains, situated agents, and the future of cognitive science. *Behavioral and Brain Sciences*, 36, 181–253.

Clark, A. (2012). Dreaming the whole cat: Generative models, predictive processing, and the enactivist conception of perceptual experience. *Mind*, 121(483), 753–771.

Clark, A. (2011). Finding the mind. *Philosophical Studies*, 152, 447–461.

Clark, A. (2009). Spreading the joy? Why the machinery of consciousness is (probably) still in the head. *Mind*, 118(472), 963–993.

Clark, A. (2008). *Supersizing the Mind*. Oxford: Oxford University Press.

Clark, A. (2006). Material symbols. *Philosophical Psychology*, 19, 291–307.

Clark, A., and Chalmers, D. (1998). The extended mind. *Analysis*, 50, 7–19.

Clark, A. (1997). *Being There*. Cambridge, MA: The MIT Press.

Cosmelli, D., and Thompson, E. (2010). Embodiment or envatment? Reflections on the bodily basis of consciousness. In J. Stewart, O. Gapenne, and E. Di Paolo (Eds), *Enaction: Towards a New Paradigm in Cognitive Science* (pp. 361–386). Cambridge, MA: The MIT Press.

Craver, C. (2007). *Explaining the Brain: Mechanisms and the Mosaic Unity of Neuroscience*. Oxford: Oxford University Press.

Dehaene, S. (2014). *Consciousness and the Brain: Deciphering How the Brain Codes Our Thoughts*. New York, NY: Viking Press.

Dehaene, S., and Changeux, J.P. (2011). Experimental and theoretical approaches to conscious processing. *Neuron*, 70(2), 200–227.

Dehaene, S. (2005). Evolution of human cortical circuits for reading and arithmetic: The neuronal recycling hypothesis. In J.R. Dehaene, M.D. Hauser, and G. Rizzolatti (Eds), *From Monkey Brain to Human Brain* (pp. 133–157). Cambridge, MA: The MIT Press.

Dehaene, S., Spelke, E., Pinel, P., Stanescu, R., and Tsivkin, S. (1999). Sources of mathematical thinking: Behavioral and brain-imaging evidence. *Science*, 284(5416), 970–974. doi: 10.1126/science.284.5416.970.

Descartes, R. (1641/1996). *Meditations on First Philosophy* (J. Cottingham, Trans.). Cambridge: Cambridge University Press.

Dewey, J. (1916). *Democracy and Education: An Introduction to the Philosophy of Education*. New York: Macmillan.

Donald, M. (1991). *Origins of the Modern Mind*. Cambridge, MA: Harvard University Press.

Drayson, Z. (2017). Modularity and the predictive mind. In T. Metzinger and W. Wiese (Eds.), *Philosophy and Predictive Processing*. Open Mind.

Dupre, J. (2018). Metaphysics of metamorphosis. https://aeon.co/essays/science-and-metaphysics-must-work-together-to-answer-lifes-deepest-questions.

Dutilh Novaes, C. (2012). Formal languages and extended cognition. In *Formal Languages in Logic: A Philosophical and Cognitive Analysis* (pp. 161–197). Cambridge: Cambridge University Press.

Edelman, G.M., and Tononi, G. (2001). *A Universe of Consciousness: How Matter Becomes Imagination*. New York, NY: Basic Books.

Feldman, H., and Friston, K. (2010). Attention, uncertainty and free energy. *Frontiers in Human Neuroscience*, 2(4), 215. doi: 10.3389/fnhum.2010.00215. eCollection 2010.

Fletcher, P., and Frith, C. (2009). Perceiving is believing: A Bayesian approach to explaining the positive symptoms of schizophrenia. *Nature Reviews Neuroscience*, 10, 48–58.

Fodor, J. (1975). *The Modularity of Mind*. Cambridge, MA: The MIT Press.

Fontanini, A., and Katz, D.B. (2008). Behavioural states, network states and sensory response variability. *Journal of Neurophysiology*, 100(3), 1160–1168.

Friston, K.J., Fortier, M., and Friedman, D.A. (2018). Of woodlice and men: A Bayesian account of cognition, life and consciousness: An interview with Karl Friston. *ALIUS Bulletin*, 2, 17–43.

Friston, K.J. (2017). Consciousness is not a thing but a process of inference. *AEON*. https://aeon.co/essays/consciousness-is-not-a-thing-but-a-process-of-inference.

Friston, K.J., Rosch, R., Parr, T., Price, C., and Bowman, H. (2017). Deep temporal models and active inference. *Neuroscience and Biobehavioral Reviews*, 77, 388–402.

Friston, K.L., Rigoli, F., Ognibene, D., Mathys, C., Fitzgerald, T., and Pezzulo, G. (2015a). Active inference and epistemic value. *Cognitive Neuroscience*, 6(4), 187–214.

Friston, K.J., Levin, M., Sengupta, D., and Pezzulo, P. (2015b). Knowing one's place: A free energy approach to pattern regulation. *The Journal of the Royal Society Interface*, 12, 20141383. doi: 10.1098/rsif.2014.1383.

Friston, K. (2013). Life as we know it. *Journal of the Royal Society, Interface*, 10, 20130475. doi: 10. 1098/rsif.2013.0475.

Friston, K., Schwartenbeck, P., Fitzgerald, T., Moutoussis, M., Behrens, T., and Dolan, R.J. (2013). The anatomy of choice: Active inference and agency. *Frontiers in Human Neuroscience*, 7. doi: 10.3389/fnhum.2013.00598.

Friston, K., Adams, R.A., Perrinet, L., and Breakspear, M. (2012a). Perception as hypotheses: Saccades as experiments. *Frontiers in Psychology*, 28(3), 151. doi: 10.3389/fpsyg.2012.00151. eCollection 2012.

Friston, K., Thornton, C., and Clark, A. (2012b). Free-energy minimization and the dark-room problem. *Frontiers in Psychology*, 3(130), 1–7.

Friston, K. (2011). Embodied inference: Or 'I think therefore I am, if I am what I think.' In W. Tschacher and C. Bergomi (Eds), *The Implications of Embodiment (Cognition and Communication)* (pp. 89–125). Exeter: Imprint Academic.

Friston, K. (2010). The free-energy principle: A unified brain theory? *Nature Reviews Neuroscience*, 11, 127–138.

Friston, K. (2009). The free-energy principle: A rough guide to the brain? *Trends in Cognitive Science*. doi: 10.1016/j.tics.2009.04.005.
Friston, K., and Stephan, K.E. (2007). Free energy and the brain. *Synthese*, 159, 417–458.
Friston, K. (2005). A theory of cortical responses. *Philosophical Transactions of the Royal Society, London B, Biological Sciences*, 360(1458), 815–836.
Frith, C., and Friston, K. (2015). A duet for one. *Consciousness and Cognition*, 390–405.
Frith, C. (2014). *Making Up the Mind: How the Brain Creates Our Mental World*. Oxford, UK: Wiley-Blackwell.
Froese, T. (2014). Steps towards an enactive account of synesthesia. *Cognitive Neuroscience*, 5(2), 126–127.
Gallagher, S. (2018). *Enactivist Interventions: Rethinking the Mind*. Oxford: Oxford University Press.
Gallagher, S. (2017). The past, present and future of time-consciousness: From Husserl to Varela and beyond. *Constructivist Foundations*, 13(1), 91–116.
Gallagher, S. (in press). New mechanisms and the enactivist concept of constitution. In M.P. Guta (Ed.), *The Metaphysics of Consciousness*. London: Routledge.
Gallagher, S., and Allen, M. (2016). Active inference, enactivism and the hermeneutics of social cognition. *Synthese* (*SI: Predictive Brains*, M. Kirchhoff (Ed.)), 1–22.
Gibbard, A. (1975). Contingent identity. *Journal of Philosophical Logic*, 4, 187–221.
Gibson, J.J. (1979). *An Ecological Approach to Vision*. New York, NY: Psychology Press.
Gibson, J.J. (1966). *The Senses Considered as Perceptual Systems*. Oxford: Houghton Mifflin.
Gillett, A.J. (2018). *A Pluralistic Approach to Distributed Cognition: Tasks, Mechanisms, and Practices*. PhD Thesis, Macquarie University, Sydney.
Gillett, C. (2007). Hyper-extending the mind? Setting Boundaries in the special sciences. *Philosophical Topics*, 35(1–2), 161–187.
Gillett, C. (2003). The metaphysics of realization, multiple realizability, and the special sciences. *The Journal of Philosophy*, 100(11), 591–603.
Gillett, C. (2002). The dimensions of realization: A critique of the standard view. *Analysis*, 64(4), 316–323.
Godfrey-Smith, P. (2016). *Other Minds: The Octopus, the Sea, and the Deep Origins of Consciousness*. New York: Farrar, Straus and Giroux.
Gray, W.D., and Fu, W.-T. (2004). Soft constraints in interactive behaviour: the case of ignoring perfect knowledge in the world for imperfect knowledge in the head. *Cognitive Science*, 28(3), 359–382.
Gregory, R. (1980). Perceptions as hypotheses. *Philosophical Transactions of the Royal Society London, Series B, Biological Sciences*, 290(1038), 181–197.
Griffiths, P., and Stotz, K. (2001). How the mind grows: A developmental perspective on the biology of cognition. *Synthese*, 122, 29–51.
Grossberg, S. (2007). Consciousness CLEARS the mind. *Neural Networks*, 20, 1040–1053.
Haken, H. (1983). *Synergetics: An Introduction: Nonequilibrium Phase Transition and Self-Organisation in Physics, Chemistry and Biology*. Berlin, Germany: Springer.

Haugeland, J. (2002). Andy Clark on cognition and representation. In H. Clapin (Ed.), *Philosophy of Mental Representation* (pp. 24–36). Oxford, UK: Oxford University Press.

Hayhoe, M.M., Shrivastava, A., Mruczek, R., and Pelz, J.B. (2003). Visual memory and motor planning in a natural task. *Journal of Vision*, 3(1), 49–63.

Heersmink, R. (2016). The metaphysics of cognitive artifacts. *Philosophical Explorations*, 19(1), 78–93.

Heersmink, R. (2014). Dimensions of integration in embedded and extended cognitive systems. *Phenomenology and the Cognitive Sciences*, 13(3), 577–598.

Helmholtz, H. (1860/1962). *Handbuch der Physiologischen Optik* (J.P.C. Southall, Ed., English Trans., Vol. 3). New York: Dover.

Hobson, J., and Friston, K.J. (2014). Consciousness, dreams, and inference: The Cartesian theatre revisited. *Journal of Consciousness Studies*, 21(1–2), 6–32.

Hoffman, E. (1989). *Lost in Translation: A Life in a New Language*. New York: Penguin Books.

Hofweber, T., and Velleman, D. (2011). How to endure. *The Philosophical Quarterly*, 61(242), 37–57.

Hohwy, J. (2017a). How to entrain your evil demon. In T. Metzinger and W. Wiese (Eds), *Philosophy and Predictive Processing* (Vol. 3). Frankfurt am Main: MIND Group. doi: 10.15502/9783958573048.

Hohwy, J. (2017b). Priors in perception: Top-down modulation, Bayesian perceptual learning rate, and prediction error minimisation. *Consciousness and Cognition*, 47, 75–85.

Hohwy, J. (2016). The self-evidencing brain. *Nous*, 50(2), 259–285.

Hohwy, J. (2015). Prediction error minimisation, mental and developmental disorder, and statistical theories of consciousness. In R. Gennaro (Ed.), *Disturbed Consciousness: New Essays on Psychopathology and Theories of Consciousness*. Cambridge, MA: The MIT Press.

Hohwy, J. (2013). *The Predictive Mind*. Oxford: Oxford University Press.

Hohwy, J. (2012). Attention and conscious perception in the hypothesis testing brain. *Frontiers in Psychology*, 1–14. doi: 10.3389/fpsyg.2012.00096.

Hohwy, J., Roepstorff, A., and Hohwy, J. (2008). Predictive coding explains binocular rivalry: An epistemological review. *Cognition*, 108(3), 687–701.

Hume, D. (1738–40/1975). *A Treatise of Human Nature* (L.A. Selby-Bigge, Ed., 2nd Edition, Revised by P.H. Nidditch). Oxford: Clarendon Press.

Hurley, S.L. (2010). The varieties of externalism. In R. Menary (Ed.), *The Extended Mind* (pp. 101–154). Cambridge, MA: The MIT Press.

Hurley, S.L. (1998). *Consciousness in Action*. Cambridge, MA: Harvard University Press.

Husserl, E. (2001). *Die Bernauer Manuskripte über das Zeitbewusstsein (1917–18) Husserliana 33*. Dordrecht: Kluwer Academic Publishers.

Hutchins, E. (2014). The cultural ecosystem of human cognition. *Philosophical Psychology*, 27, 34–49.

Hutchins, E. (2011). Enculturating the supersized mind. *Philosophical Studies*, 152, 437–446.

Hutchins, E. (2008). The role of cultural practices in the emergence of modern human intelligence. *Philosophical Transaction of the Royal Society*, 363, 2011–2019.

Hutchins, E. (1995). *Cognition in the Wild*. Cambridge: The MIT Press.

Hutto, D.D., and Myin, E. (2017). *Evolving Enactivism: Basic Minds Meet Content*. Cambridge, MA: The MIT Press.

Hutto, D.D., and Myin, E. (2013). *Radicalising Enactivism*. Cambridge, MA: The MIT Press.

Hutto, D.D., Kirchhoff, M.D., and Myin, E. (2014). Extensive enactivism: Why keep it all in? *Frontiers in Human Neuroscience*, 1–11. doi: 10.3389/fnhum.2014.00706.

Hutto, D.D. (2005). Knowing what? Radical versus conservative enactivism. *Phenomenology and the Cognitive Sciences*, 4(4), 389–405.

James, W. (1890/1981). *The Principles of Psychology*. Cambridge, MA: Harvard University Press.

Kelso, S. (1995). *Dynamic Patterns*. Cambridge: The MIT Press.

Kim, J. (1998). *Mind in a Physical World*. Cambridge, MA: The MIT Press.

Kirchhoff, M.D., and Kiverstein, J. (under review). Nested Markov blankets and the boundaries of mind. *Nous*.

Kirchhoff, M.D., and Robertson, I. (2018). Enactivism and predictive processing: a non-representational view. *Philosophical Explorations*. doi: 10.1080/13869795.2018.1477983.

Kirchhoff, M.D. (2018). Hierarchical Markov blankets and adaptive active inference. *Physics of Life Reviews*, 1–2. https://doi.org/10.1016/j.plrev.2017.09.001.

Kirchhoff, M.D., Parr, T., Palacios, E., Friston, K.J., and Kiverstein, J. (2018). The Markov blankets of life: Active inference, autonomy and the free energy principle. *The Journal of the Royal Society Interface*, 15, 20170792. http://dx.doi.org/10.1098/rsif.2017.0792.

Kirchhoff, M.D. (2017). Predictive processing, perceiving and imagining: Is to perceive to imagine, or something close to it? *Philosophical Studies*, 175(3), 751–767.

Kirchhoff, M.D., and Froese, T. (2017). Where there is life there is mind: In support of a strong life-mind continuity thesis. *Entropy*, 19(169). doi: 10.3390/e19040169.

Kirchhoff, M.D., and Meyer, R. (2017). Breaking explanatory boundaries: Flexible borders and plastic minds. *Phenomenology and the Cognitive Sciences*, 1–20. https://doi.org/10.1007/s11097-017-9536-9.

Kirchhoff, M.D. (2016). Autopoiesis, free energy, and the life-mind continuity thesis. *Synthese*, 1–22. doi: 10.1007/s11229-016-1100-6.

Kirchhoff, M.D., and Hutto, D.D. (2016). Never mind the gap: Neurophenomenology, radical enactivism, and the hard problem of consciousness. *Constructivist Foundations*, 11(2), 346–353.

Kirchhoff, M.D. (2015). Cognitive assembly: Towards a diachronic notion of composition. *Phenomenology and the Cognitive Sciences*, 14(1), 33–53.

Kirchhoff, M.D. (2015a). Extended cognition & the causal-constitutive fallacy: In search for a diachronic and dynamical conception of constitution. *Philosophy and Phenomenological Research*, 90(2), 320–360.

Kirchhoff, M.D. (2015b). Experiential fantasies, prediction and enactive minds. *Journal of Consciousness Studies*, 22(3–4), 68–92.

Kirchhoff, M.D. (2015c). Species of realization and the free energy principle. *Australasian Journal of Philosophy*, 93(4), 706–723.

Kirchhoff, M.D. (2014). Extended cognition & constitution: Re-evaluating the constitutive claim of extended cognition. *Philosophical Psychology*, 27(2), 258–283.

Kirchhoff, M.D. (2012). Extended cognition and fixed properties: Step to a third-wave version of extended cognition. *Phenomenology and the Cognitive Sciences*, 11, 287–308.

Kirchhoff, M.D., and Newsome, W. (2012). Distributed cognitive agency in virtue epistemology. *Philosophical Explorations*, 15(2), 165–180.

Kirmayer, L. (2018). Ontologies of life: From thermodynamics to teleonomics comment on 'Answering Schrödinger's question: A free-energy formulation' by Maxwell James Désormeau Ramstead et al. *Physics of Life Reviews*. https://doi.org/10.1016/j.plrev.2017.11.022.

Kirsh, D., and Maglio, P. (1994). On distinguishing epistemic from pragmatic action. *Cognitive Science*, 18(4), 513–549.

Kiverstein, J. (2018). The free energy self: An ecological-enactive interpretation. In A. Ciaunica (Ed.), *Topoi: (SI: The Relational Self)*.

Kiverstein, J. (2016). The interdependence of embodied cognition and consciousness. *Journal of Consciousness Studies* 23(5–6): 105–137.

Kiverstein, J., Miller, M., and Rietveld, E. (2017). The feeling of grip: Novelty, error dynamics and the predictive brain. *Synthese*, 1–23. doi: https://doi.org/10.1007/s11229-017-1583-9

Kiverstein, J., Farina, M., and Clark, A. (2015). Substituting the senses. In M. Matthen (Ed.), *The Oxford Handbook of the Philosophy of Perception*. New York, NY: Oxford University Press.

Knill, D.C., and Pouget, A. (2004). The Bayesian Brain: The role of uncertainty in neural coding and computation. *Trends in Neuroscience*, 27(12), 712–719.

Koschmieder, E.L. (1993). *Bénard Cells and Taylor Vortices*. Cambridge: Cambridge University Press.

Lamme, V. (2015). The crack of dawn: Perceptual functions and the neural mechanisms that mark the transition from unconscious processing to conscious vision. In T. Metzinger and J.M. Windt (Eds.), *Open Mind 22* (pp. 1–34). Frankfurt am Main: MIND Group. doi: 10.15502/9783958570092.

Latour, B. (1996). Cogito ergo sumus! A review of Ed Hutchins *Cognition in the Wild. Mind, Culture, and Activity: An International Journal*, 3(1), 54–63.

Lave, J., and Wegner, E. (1991). *Situated Learning: Legitimate Peripheral Participation*. Cambridge: Cambridge University Press.

Lawson, R.P., Rees, G., and Friston, K. (2014). An aberrant precision account of autism. *Frontiers in Human Neuroscience*, 8, 302.

Levine, J. (1983). Materialism and qualia: The explanatory gap. *Pacific Philosophical Quarterly*, 64, 354–361.

Lloyd, D. (2017). Protention and predictive processing: The wave of the future. *Constructivist Foundations*, 13(1), 98–99.

Lupyan, G. (2015). Cognitive penetrability of perception in the age of prediction: Predictive systems are penetrable systems. *Review of Philosophy and Psychology*, 6(4), 547–569.

Lupyan, G., and Ward, E.J. (2013). Language can boost otherwise unseen objects into visual awareness. *Proceedings of the National Academy of Sciences*, 110(35), 14196–14201.

Macpherson, F. (2017). The relationship between cognitive penetration and predictive coding. *Consciousness and Cognition*, 47, 6–16.

Madary, M. (2016). *Visual Phenomenology*. Cambridge, MA: The MIT Press.

Malafouris, L. (2010). Metaplasticity and the human becoming: Principles of neuroarchaeology. *Journal of Anthropological Sciences*, 88, 49–72.

Malafouris, L. (2004). The cognitive basis of material engagement: Where brain, body and culture conflate. In E. DeMarrais, C. Gosden, and C. Renfrew (Eds), *Rethinking Materiality: The Engagement of Mind with the Material World* (pp. 53–62). Cambridge: McDonald Institute Monographs.

Melloni, L., and Singer, W. (2010). Distinct characteristics of conscious experience are met by large-scale neuronal synchronisation. In E.K. Perry, D. Collerton, F.E.N. LeBeau, and H. Ashton (Eds.), *New Horizons in the Neuroscience of Consciousness* (pp. 17–28). Amsterdam, NL: John Benjamins.

Menary, R. (2015). Mathematical cognition: A case of enculturation. In T. Metzinger and J.M. Windt (Eds), *Open MIND: 25(T)*. Frankfurt am Main: MIND Group. doi: 10.15502/9783958570818.

Menary, R., and Kirchhoff, M.D. (2014). Cognitive transformations and extended expertise. *Educational Philosophy and Theory*, 46(6), 1–14.

Menary, R. (2010). Cognitive integration and the extended mind. In R. Menary (Ed.), *The Extended Mind* (pp. 227–243). Cambridge, MA: The MIT Press.

Menary, R. (2007). *Cognitive Integration: Mind and Cognition Unbounded*. Basingstoke: Palgrave Macmillan.

Merckelbach, H., and van de Ven, V. (2001). Another White Christmas: Fantasy proneness and reports of 'hallucinatory experiences' in undergraduate students. *Journal of Behavioural Therapy and Experimental Psychiatry*, 32, 137–144.

Mirza, B.M., Adams, R.A., Mathys, C.D., and Friston, K.J. (2016). Scene construction, visual foraging, active inference. *Frontiers in Computer Neuroscience*, 10, 56. doi: 10.3389/fncom.2016.00056.

Murphy, K.P. (2012). *Machine Learning: A Probabilistic Perspective*. Cambridge, MA: The MIT Press.

Nagy, E. (2008). Innate intersubjectivity: newborn's sensitivity to communication disturbance. *Developmental Psychology*, 44(6), 1779–84.

Noë, A. (2016). Sensations and situations: A sensorimotor integrationist approach. *Journal of Consciousness Studies*, 23(5–6), 66–79.

Noë, A. (2012). *The Varieties of Presence*. Cambridge, MA: Harvard University Press.

Noë, A. (2009). *Out of Our Heads: Why You Are Not Your Brain, and Other Lessons from the Biology of Consciousness*. New York: Hill and Wang.

Nöe, A. (2006). Experience of the world in time. *Analysis* 66(1), 26–32.

Noë, A. (2004). *Action in Perception*. Cambridge, MA: The MIT Press.

O'Regan, J.K., and Noë, A. (2001). A sensorimotor account of vision and visual consciousness. *Behavioural and Brain Sciences*, 24(5), 973–1031.

Palacios, E., Razi, A., Parr, T., Kirchhoff, M.D., and Friston, K. (2017). Biological self-organisation and Markov blankets. *BioRxiv*, 227181. http://dx.doi.org/10.1101/227181.

Palmer, C.J., Lawson, R.P., and Howly, J. (2017). Bayesian approaches to autism: Towards volatility, action and behaviour. *Psychological Bulletin*. Advance online publication. http://dx.doi.org/10.1037/bul0000097.

Parr, T., Rees, G., and Friston, K.J. (2018). Computational neuropsychology and Bayesian inference. *Frontiers in Human Neuroscience*, 12(61). doi: 10.3389/fnhum.2018.00061.

Paton, B., Skewes, J., Frith, C., and Hohwy, J. (2013). Skull-bound perception and precision optimization through culture. *Behavioral and Brain Sciences*, 36(4), 222.

Pearl, J. (1988). *Probabilistic Reasoning in Intelligent Systems: Networks of Plausible Inference*. San Francisco, CA: Morgan Kaufmann Publishers.

Pellicano, E., and Burr, D. (2012). When the world becomes 'too real': A Bayesian explanation of autistic perception. *Trends in Cognitive Science*, 16(10), 504–510.

Pezzulo, G., and Levin, M. (2018). Embodying Markov blankets comment on 'Answering Schrödinger's question: A free-energy formulation" by Maxwell James Désormeau Ramstead et al. *Physics of Life Reviews*. https://doi.org/10.1016/j.plrev.2017.11.0201571-0645.

Powers, W.T. (1973). *Behaviour: The Control of Perception*. New Canaan, CT: Benchmark Publishers.

Prigogine, I., and Nicolis, G. (1971). Biological order, structure and instabilities. *Quarterly Reviews of Biophysics*, 4, 107–148. doi: 10.1017/S0033583500000615.

Prinz, J. (2008). Is consciousness embodied? In M. Aydede and P. Robbins (Eds.), *The Cambridge Handbook of Situated Cognition* (pp. 419–437). Cambridge, UK: Cambridge University Press.

Putnam, H. (1975). The meaning of 'meaning.' In K. Gunderson (Ed.), *Language, Mind, and Knowledge*. Minneapolis: University of Minneapolis Press.

Rabinovich, M.I., Friston, K.J., and Varona, P. (2012). *Principles of Brain Dynamics: Global State Interactions*. Cambridge, MA: The MIT Press.

Ramstead, M., Badcock, P., and Friston, K.J. (2017). Answering Schrödinger's question: A free-energy formulation. *Physics of Life Reviews*, 1–29. doi: 10.1016/j.plrev.2017.09.001.

Ramstead, M.J.D., Kirchhoff, M.D., and Friston, K. (under review). *A tale of two densities: active inference and anti-realism about Bayesian cognitive science*. Adaptive Behavior.

Rao, R.P., and Ballard, D.H. (1999). Predictive coding in the visual cortex: A functional interpretation of some extra-classical visual-field effects. *Nature Neuroscience*, 2, 79–87.

Ratcliffe, M. (2017). *Real Hallucinations: Psychiatric Illness, Intentionality, and the Interpersonal World*. Cambridge, MA: The MIT Press.

Revonsuo, A. (2015). Hard to see the problem? *Journal of Consciousness Studies*, 22(3–4), 52–67.

Rickles, D., Hawe, P., and Shiell, A. (2007). A simple guide to chaos and complexity. *Journal of Epidemiology and Community Health*, 69, 933–937.

Rietveld, E., and Kiverstein, J. (2014). A rich landscape of affordances. *Ecological Psychology*, 26(4), 325–352.

Roberts, T. (2010). Understanding 'sensorimotor understanding.' *Phenomenology and the Cognitive Sciences*, 9(1), 101–111.

Roepstorff, A., Niewöhner, J., and Beck, S. (2010). Enculturating brains through patterned practices. *Neural Networks*, 23, 1051–1059.

Roepstorff, A. (2008). Things to think with: Words and objects as material symbols. *Philosophical Transactions of the Royal Society, London B*, 363, 2049–2054.

Roepstorff, A., and Frith, C. (2004). What's at the top in the top-down control of action? Script-sharing and 'top-top' control of action in cognitive experiments. *Psychological Research*, 68, 189–198.

Rorty, R. (1979). *Philosophy and the Mirror of Nature*. Princeton, NJ: Princeton University Press.

Rowlands, M. (2010). *The New Science of Mind*. Cambridge, MA: The MIT Press.

Rowlands, M. (2009). Enactivism and the extended mind. *Topoi*, 28, 53–62.

Rowlands, M. (2007). Understanding the 'active' in enactive. *Phenomenology and the Cognitive Sciences*, 6(4), 427–443.

Rowlands, M. (1999). *The Body in Mind: Understanding Cognitive Processes*. Cambridge: Cambridge University Press.

Rubin, E. (1921). *Visuell Wahrgenommene Figuren*. Copenhagen: Gyldendal.

Rumelhart, D., McClelland, J.L., and Hinton, G.E. (1986). The appeal of parallel distributed processing. In J.L. McClelland, D. Rumelhart and the PDP Research Group (Ed's) *Parallel Distributed Processing: explorations in the microstructure of cognition*. Cambridge, MA: The MIT Press.

Rupert, R. (2015). Embodiment, consciousness, and neurophenomenology: Embodied cognitive science puts the (first) person in its place. *Journal of Consciousness Studies*, 22(3–4), 148–180.

Rupert, R. (2009). *Cognitive Systems and the Extended Mind*. Oxford: Oxford University Press.

Rupert, R. (2004). Challenges to the hypothesis of extended cognition. *Journal of Philosophy*, 9, 625–636.

Sass, L. (1994). *The Paradoxes of Delusion: Wittgenstein, Schreber, and the Schizophrenic Mind*. Cornell, NY: Cornell University Press.

Seth, A.K. (2015a). The cybernetic brain: From interoceptive inference to sensorimotor contingencies. In T. Metzinger and J.M. Windt (Eds), *Open MIND: 35(T)*. Frankfurt am Main, Germany: MIND Group.

Seth, A.K. (2015b). Presence, objecthood, and the phenomenology of predictive perception. *Cognitive Neuroscience*, 6(2–3), 111–117.

Seth, A.K. (2014). A predictive processing theory of sensorimotor contingencies: Explaining the puzzle of perceptual presence and its absence in synesthesia. *Cognitive Neuroscience*, 5(2), 97–118.

Seth, A.K., Suziki, K.S., and Critchley, H.D. (2012). An interoceptive predictive coding model of conscious presence. *Frontiers in Psychology: Consciousness Research*, 2, 395. doi: 10.3389/fpsyg.2011.00395.

Shankar, M.U., Levitan, C., and Spence, C. (2010). Grape expectations: The role of cognitive influences in colour-flavour interactions. *Consciousness & Cognition*, 19, 380–390.

Shapiro, L. (2011). *Embodied Cognition*. Oxford: Routledge.

Spinoza, B. (1677/1996). *Ethics* (E.M. Curley, Trans.). London, UK: Penguin Books.

Sporns, O. (2011). *Networks of the Brain*. Cambridge, MA: The MIT Press.

Sprevak, M. (2009). Extended cognition and functionalism. *The Journal of Philosophy*, 106(9), 503–527.

Sterelny, K. (2012). *The Evolved Apprentice*. Cambridge, MA: The MIT Press.

Sterelny, K. (2010). Minds: Extended or scaffolded? *Phenomenology and the Cognitive Sciences*, 9(4), 465–481.

Stotz, K. (2010). Human nature and cognitive-developmental niche construction. *Phenomenology and the Cognitive Sciences*, 9(4), 483–501.

Sur, M., Angelucci, A., and Sharma, J. (1999). Rewiring cortex: The role of patterned activity in development and plasticity of neocortical circuits. *Developmental Neurobiology*, 41(1), 33–43.

Sutton, J. (2010). Exograms and interdisciplinarity: History, the extended mind, and the civilizing process. In R. Menary (Ed.), *The Extended Mind* (pp. 189–225). Cambridge, MA: The MIT Press.

Sutton, J., Harris, C.B., Keil, P.G., and Barnier, A.J. (2010). The psychology of memory, extended cognition, and socially distributed remembering. *Phenomenology and the Cognitive Sciences*, 9(4), 521–560.

Sutton, J. (2008). Material agency, skills and history: Distributed cognition and the archaeology of memory. In C. Knappett and L. Malafouris (Eds), *Material Agency* (pp. 37–55). Springer.

Theiner, G., Allen, C., and Goldstone, R.L. (2010). Recognizing group cognition. *Cognitive Systems Research*, 11, 378–395.

Thierry, G., Athanasopoulos, P., Wiggett, A., Dering, B., and Kuipers, J.R. (2009). Unconscious effects of language-specific terminology on preattentive color perception. *Proceedings of the National Academy of Sciences*, 106(11), 4567–4570.

Thiese, N.D., and Kafatos, M. (2013). Complementarity in biological systems: A complexity view. *Complexity*, 18, 11–20.

Thompson, E. (2007). *Mind in Life: Biology, Phenomenology, and the Sciences of Mind*. Cambridge, MA: Harvard University Press.

Thompson, E., and Varela, F. (2001). Radical embodiment, neural dynamics and consciousness. *Trends in Cognitive Science*, 5(10), 418–425.

Thurston, J.B., and Carraher, R.G. (1966). *Optical Illusions and the Visual Arts*. New York: Van Nostrand Reinhold.

Tribble, E.B. (2005). Distributing cognition in the Globe. *Shakespeare Quarterly*, 56(2), 135–155.

Turner, S. (2000). *The Extended Organism*. Harvard: Harvard University Press.

Tye, M. (2017). Qualia. *Stanford Encyclopedia of Philosophy*. https://plato.stanford.edu/entries/qualia/.

Tylen, K., Wallentin, M., and Roepstorff, A. (2009). Say it with flowers! An fMRI study of object mediated communication. *Brain and Language*, 108, 159–166.

Van Gelder, T., and Port, R. (1995). It's about time: An overview of the dynamical approach to cognition. In R. Port and T. van Gelder (Eds), *Mind as Motion: Explorations in the Dynamics of Cognition* (pp. 1–44). Cambridge, MA: The MIT Press.

Varela, F.J., Lachaux, J.-P., Rodriguez, E., and Martinerie, J. (2001). The brainweb: Phase synchronization and large-scale integration. *Nature Reviews Neuroscience*, 2, 229–239.

Varela, F.J. (1999a). The specious present: A neurophenomenology of time consciousness. In J. Petitot, F.J. Varela, B. Pachoud, and J.-M. Roy (Eds), *Naturalizing*

Phenomenology: Issues in Contemporary Phenomenology and Cognitive Science (pp. 266–329). Stanford: Stanford University Press. http://cepa.info/2081.

Varela, F.J. (1999b). Present time-consciousness. *Journal of Consciousness Studies*, 6(2–3), 111–140.

Varela, F. (1996). Neurophenomenology: A methodological remedy for the hard problem. *Journal of Consciousness Studies*, 3(4), 330–349.

Vold, K. (2015). The parity argument for extended consciousness. *Journal of Consciousness Studies*, 22(3–4), 16–33.

Vygotsky, L.S. (1978). *Mind in Society: The Development of Higher Psychological Processes*. Cambridge, MA: Harvard University Press.

Ward, D. (2012). Enjoying the spread: Conscious externalism reconsidered. *Mind*, 121(483), 731–751.

Ward, J. (2013). Synesthesia. *Annual Review of Psychology*, 64, 49–75.

Wasserman, R. (2004). The constitution question. *Nous*, 38(4), 693–710.

Wertsch, J.V. (1985). *Vygotsky and the Social Formation of Mind*. Cambridge, MA: Harvard University Press.

Wexler, B. (2008). *Brain and Culture: Neurobiology, Ideology, and Social Change*. Cambridge, MA: The MIT Press.

Wheeler, M. (2015a). Extended consciousness: An interim report. *Southern Journal of Philosophy*, 53(1), 155–175.

Wheeler, M. (2015b). Not what it's like but where it's like: Phenomenal consciousness, sensory substitution and the extended mind. *Journal of Consciousness Studies*, 22(3–4), 129–147.

Wheeler, M. (2010). In defense of extended functionalism. In R. Menary (Ed.), *The Extended Mind* (pp. 245–270). Cambridge, MA: The MIT Press.

Williams, D. (2017). Predictive processing and the representation wars. *Minds and Machines*, 28(1), 141–172.

Wilson, R.A., and Clark, A. (2009). How to situate cognition: Letting nature take its course. In P. Robbins and M. Aydede (Eds), *The Cambridge Handbook of Situated Cognition* (pp. 55–77). Cambridge: Cambridge University Press.

Wilson, R.A. (2004). *Boundaries of the Mind*. New York: Cambridge University Press.

Wilson, R.A. (1994). Wide computationalism. *Mind*, 103(411), 351–372.

Windt, J. (2017). Predictive brains, dreaming selves, sleeping bodies: The analysis of dream movement can inform a theory of self-and world-simulation in dreams. *Synthese*. doi: 10.1007/s11229-017-1525-6.

Index

For Product Safety Concerns and Information please contact our EU representative GPSR@taylorandfrancis.com Taylor & Francis Verlag GmbH, Kaufingerstraße 24, 80331 München, Germany